TEXT BOOK OF ADVANCED POLARIZATION

TEXT BOOK OF ADVANCED POLARIZATION

K.P. Das

CYBER TECH PUBLICATIONS
4264/3, Ansari Road, Daryaganj, New Delhi-110002 (India)
Ph.: 011-23244078, 011-43559448 Fax: 011-23280028
E-Mail: cyberpublicationsdelhi@yahoo.com
Website: www.cybertechpublications.com

TEXT BOOK OF ADVANCED POLARIZATION

K.P. Das

First Edition 2010

Asian Offset Printers

Published by :

G.S. Rawat for **Cyber Tech Publications**
4264/3, Ansari Road, Daryaganj, New Delhi-110002 (India)
Ph.: 011-23244078, 011-43559448 Fax: 011-23280028
E-Mail: cyberpublicationsdelhi@yahoo.com
Website: www.cybertechpublications.com

PREFACE

This books "Text Book of Advanced Polarization" has been specially written to the requirements syllabus of the students of all Indian Universities. The proofs of various theorems and examples has been given with minute details. Each chapter of this book contains complete theory and fairly large number of solved examples. The subject matter has been discussed in such a simple way that the students will find no difficulty to understand it.

Sufficient problems has also been selected from various universities examination papers. and the author shall be thankful to the readers who point out errors and omissions which inspite of all care might have been there.

The author will feel amply rewarded if the book serve the purpose for which it is meant. The student suggestion some errors for the improvement of the book are always welcome.

—Author

CONTENTS

1

Electromotive Force of Polarization

INTRODUCTION

If a difference of potential of about 1 volt is applied to two platinum electrodes immersed in a concentrated solution of HCl, the current which passes at first will steadily diminish and ultimately will become zero. The cessation of the current is due to accumulation of H_2 on anode and Cl_2 on cathode. These two gases set out an opposing electromotive force known as electromotive force of polarization. If E is the e.m.f. of the external source. R is the resistance of the whole circuit and I, the current flowing through the cell, it is found that in general, I is not equal in magnitude to E/R but may be appreciably less. In fact, the chemical changes that occur at the electrode have converted the electrolytic cell into a voltaic cell with an e.m.f., e the direction of which is opposite to that of E, then

$$I = \frac{E - e}{R} \qquad ...(1)$$

where e is counter or back e.m.f. of electrolytic cell, or e.m.f. of polarization. The magnitude of polarization depends on :

(i) Temperature

(ii) The nature of the electrodes

(iii) The nature of the solution and

(iv) The current densities at each electrode.

According to the convention followed earlier, e.m.f., e, can be written in terms of the electromotive forces, e_c and e_a at the cathode and anode, respectively.

$$e = e_c - e_a \qquad ...(2)$$

From equation (1) $E = (e_c - e_a) + RI$

If π_c and π_a are the potential differences at cathode and anode, respectively, during electrolysis, then according to conventions,

$$\pi_c = -c_c, \pi_a = -e_a$$

Therefore, $e = \pi_a - \pi_c$

OPTICAL ACTIVITY

When a beam of plane polarised light propagates through a quartz crystal along the optic axis, the plane of polarisation steadily turns about the direction of the beam. The optical rotation can be detected as follows. If two Polaroid sheets or Nicol prisms are held in crossed configuration and if a beam of unpolarized light is viewed through them, the field of view appears to be completely dark.

Now let a quartz crystal, cut with its faces perpendicular to the optic axis, be inserted between the polarisers such that light is incident normally on the crystal. The field of view now appears lit up indicating that the light is not cut off by the analyser. In order to cut off the transmitted light, we find that the analyser is to be rotated through a certain angle.

The experiment establishes that the plane polarised light produced by the polarizer remains plane polarised while passing through the quartz crystal but the plane of polarisation is rotated through an angle. This angle is the angle through which the analyser is rotated in order to cut off the light totally.

The ability to rotate the plane of polarisation of plane polarised light by certain substances is called *optical activity*. Substances, which have the ability to rotate the plane of the polarised light passing through them, are called optically active substances. Quartz and cinnabar are examples of *optically active* crystals while aqueous solutions of sugar, tartaric acid are optically active solution and liquid.

Optically active substances are classified into two types.

(i) *Dextrorotatory substances :* Substances which rotate the plane of polarization of the light towards the right are known as right-handed or dextrorotatory.

(ii) *Laevorotatory substances :* Substances which rotate the plane of polarization of the light towards the left are known as left-handed or laevorotatory.

Fresnel's Explanation of Optical Rotation

A linearly polarised light can be considered as a resultant of two circularly polarised vibrations rotating in opposite directions with the same angular velocity. Fresnel assumed that plane-polarised light on entering a crystal along the optic axis is resolved into two circularly polarised vibrations rotating in opposite directions with the same angular frequency. In a crystal like calcite, the two circularly polarised vibrations travel with the same angular velocity.

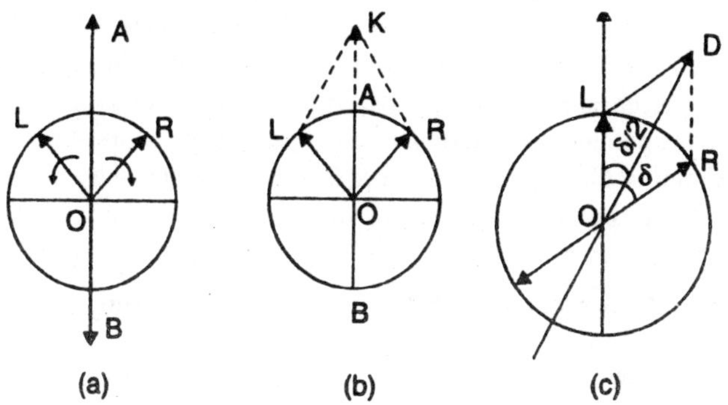

Fig. 1 : (a) A linearly polarized light wave can be regarded as a resultant of a right regularly polarized wave and a left circularly polarized wave. (b) In optically inactive materials, the vibrations corresponding to the circularly polarized waves rotate with equal angular velocity, (c) In optically active substances, the vibrations rotate with different angular velocities.

In circularly polarised vector rotating in the anticlockwise direction and OR is the circularly polarised vector rotating in the clockwise direction. The resultant of OR and OL is the vector OA. According to Fresnel, when linearly polarised light enters a crystal of calcite along the optic axis, the circularly polarised vibrations rotating in opposite directions have the same velocity.

The resultant vibration will be along OK. Therefore, crystals like calcite do not rotate the plane of vibration. In case of quartz, the linearly polarised light, the component having clockwise rotation travels faster than the anticlockwise component. When the components emerge out of the crystal, they are at an angle 5. The resultant of these two vectors OR and OL is now along OD.

Before entering the crystal, the plane of vibration is along OA and after emerging from the crystal, it is along OD. The plane of vibration has rotated through an angle δ/2. The angle through which the plane of vibration is rotated depends on the thickness of the crystal.

DECOMPOSITION VOLTAGE OR DECOMPOSITION POTENTIAL

We have already discussed that in the phenomenon of polarisation the products of electrolysis exert a back E.M.F. which reduces the actual E.M.F. of the cell. Thus, the electrolysis of an electrolyte is only possible when this back E.M F. is overcome. For example, when two smooth platinum electrodes are placed in a dilute solution of H_2SO_2 and low voltage is applied, the electrolysis takes place but it stops immediately. This is due to the fact that the back E.M.F. is greater than the original E.M.F. If the applied E.M.F. is gradually increased, the electrolysis starts and continues even though the applied E.M.F. is increased no more.

Thus the applied voltage (or E.M F.) *which is just sufficient to overcome the back E.M.F. due to polarisation and also to bring about electrolysis of an electrolyte without any hindrance is known as decomposition potential.* The voltage is seen to increase rapidly at first while the current remains very small but at a certain point (D) there is a sudden rise in current and the voltage changes by only a small amount. The voltage at which there is a sudden increase in current density is known as decomposition voltage or decomposition position voltage or decomposition potential.

Table 1

Electrolyte	Decomposition voltage	Electrolyte	Decomposition voltage
$AgNO_3$	0.80	HNO_3	1.69
HCl	1.31	NH_4OH	1.74
$CuSO_4$	1.20	$NiCl_2$	1.85
$Pb(NO_3)_2$	1.53	$CaSO_4$	1.92
KOH	1.67	$NiSO_4$	2.09
H_2SO_4	167	$ZnSO_4$	2.55

Factors affecting decomposition potential : The decomposition potential is independent of the shape of cell and the distance between the electrodes. However, it depends upon the nature of the solution and the material of the electrode.

Measurement of decomposition potential or voltage : This can be measured in two ways :

First Method : In this method, the external E.M.F. applied to the electrodes is gradually increased and the point is, noted at which there is a sudden increase in the current. This point gives the measure of decomposition potential. E and E' are the two platinum electrodes dipped in an electrolyte. S is a stirrer for stirring the Solution. The two electrodes E and E' are connected to a voltmeter V through ammeter M to measure the strength of the current. B is the external source of E.M.F. The voltage applied to the cell can be regulated with the help of a variable resistance C, and this voltage in turn can be measured with the help of a voltmeter V. In order to measure the decomposition potential, a series of current v/s applied voltage readings are taken. These readings may be taken above and below the decomposition potential.

In the beginning sufficient resistance is applied to C to produce only a very small potential and both voltmeter and ammeter readings are noted. Then, the resistance is reduced on C and again, both voltmeter and ammeter readings are noted. This procedure is continued till an appreciable current is passing through the circuit and the electrolysis in the cell is visible. The curve AB represents the current below the decomposition potential and the current is very small. After the points B and E. There is a sudden increase in the current and hence the current takes a sharp turn upwards. The point of intersection of BD and ED is the decomposition voltage.

Second Method : The applied voltage is increased gradually until bubbles of gas or deposition of metal on an electrode can be detected by viewing through a microscope. This method is not preferred as compared with the first method.

Discharge or Deposition Potential

Decomposition voltage can be split up into two parts. Each part represents that voltage which should be applied to each of the two electrodes to bring about the decomposition of the electrolyte in use. Let us consider a reversible cell having two electrodes, viz; one electrode consisting of a zinc rod dipped into a solution of M–$ZnBr_2$ and other electrode consisting of a platinum wire dipped in pure liquid bromine.

The two electrodes are connected through a salt bridge to eliminate liquid-liquid junction potential. Suppose this cell may be represented as

$$Zn \mid M\text{–}ZnBr_2\ (aq) \parallel Br_2\ (l),\ Pt$$

The half-cell reactions may be put as

$$Zn \rightarrow Za^2 + 2e^- \text{ Oxidation potential} = 0.76 \text{ V}$$

$$Br_2 + 2e^- \rightarrow 2Br^-; \text{ Oxidation potential} = 1.06 \text{ V}$$

The complete cell reaction is, therefore, as follows :

$$Zn + Br_2 \rightarrow Zn^{2+} + 2B^-$$

The E.M.F.of the cell may be given by

$$E = 0.76 - (-1.06) = 1.82\text{V}$$

In the continuous decomposition of a molar solution of zinc bromide, two observations are made :

(i) The potential at the cathode should be Just greater than 0.76 volt. It means that zinc ions will only be discharged on the zinc electrode (cathode) if the potential on the zinc electrode just exceeds 0.76 volt. The potential of zinc electrode is known as discharge or deposition potential of cation.

(ii) The potential at the anode should be greater than 1.06. It means that bromine ions will only be discharged as bromine on the platinum electrode if the potential of the Pt anode just exceeds 1.06 volt. This potential of Pt electrode is known as discharge or deposition potential of Br^- anion.

From the above discussion, it follows that "*the discharge or deposition potential of an ion is defined as the potential which should be applied to electrode so as to bring about a continuous discharge or deposition of the ion at that electrode.*" The discharge or deposition potential of an ion depends upon its concentration *i.e.*, the greater the concentration of the ion the lesser will be its discharge potential.

Applications of Decomposition Voltage

(i) If a solution is having a number of ions and is subjected to electrolysis, then the ions will be discharged in the order of their increasing decomposition voltages. This fact is utilised in the separation of cations. For example, the decomposition potentials of $Cd(NO_3)_2$, $Pb(NO_3)_2$ and $AgNO_3$ are 2.3, 1.8 and 0.9 volts respectively. If an E.M.F. greater than one volt is applied, silver is only deposited. Now if the E.M.F. is increased to 1.9 volts, only lead will get deposited. Now if the E.M.F. of about 2.3 volts is applied, only cadmium will get deposited.

(ii) The idea of decomposition potential is of great importance in electroplating. For example, if in nickel plating, the activity exceeds a certain limit, the hydrogen will be evolved and deposition will not be firm and smooth.

(iii) In the manufacture of brass by electrolytic method from zinc and copper, the idea of decomposition potential plays an important role.

FRESNEL'S RHOMB

Fresnel designed a rhomb of glass whose angles are 54° and 126°. Its functioning is based on the fact that a phase difference of n $\pi/4$ is introduced between the component vibrations when light is totally internally reflected back at glass - air- interface when the angle of incidence is 54°. A ray of light enters normally at one end of the rhomb and is totally internally reflected at the point B along BC. The angle of incidence at B is 54°, which is more than the critical angle of glass. Let the incident light be plane polarised and let the vibrations make an angle of 45° with the plane of incidence. Its components (i) parallel to the plane of incidence and (ii) perpendicular to the plane of incidence are equal. These components after reflection at the point B undergo a phase difference of $\pi/4$ or a path difference of $\lambda/8$.

A further phase difference of $\pi/4$ or a path difference of $\lambda/8$ is introduced between the components when the ray BC is totally internally reflected back along CD. Therefore, the final emergent ray DE has two components, vibrating at right angles to each other and they have a path difference of $\lambda/4$. Therefore, the emergent ray is circularly polarised. If the light entering the Fresnel's rhomb is circularly polarised, a further path difference of $\lambda/4$ is introduced between the component vibrations. The total path difference between the component vibrations is $\lambda/2$. Therefore, the emergent light is plane polarised and its vibrations make an angle of 45° with the plane of incidence.

When an elliptically polarised light is passed through a Fresnel's rhomb, a further path difference of $\lambda/4$ is introduced between the component vibrations. The total path difference between the component vibrations is $\lambda/2$ and the emergent light is plane polarised.

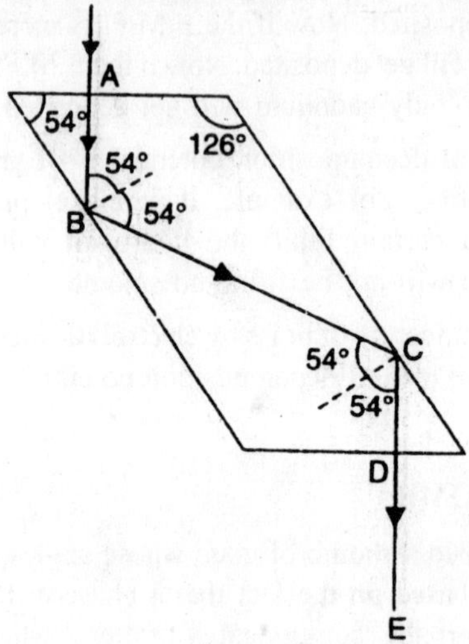

Fig. 2

Thus, Fresnel's rhomb works very much similar to a quarter wave plate. A quarter wave plate is used for a particular wavelength light whereas a Fresnel's rhomb can be used for light of all wavelengths.

DOUBLE IMAGE POLARIZING PRISMS

Nicol prism cannot be used with ultraviolet light on account of the canada balsam layer which absorbs UV rays. Sometimes it is also desirable to have both the ordinary and extraordinary rays widely separated. For this purpose, two prisms, namely Rochon prism and Wollaston prism, are used.

1. Rochon Prism

It consists of two prisms ABC and BCD (of quartz or calcite) cut with their optic axes as shown in Fig. 3. The prism ABC is cut such that the optic axis is parallel to the face AB and the incident light. The prism BCD has the optic axis perpendicular to the plane of incidence. Light incident normally on the face AC of the prism passes without deviation up to the boundary BC. In the prism BCD, the ordinary ray passes without deviation. If the prisms are made of quartz, the extraordinary ray is deviated as shown in Fig. 3.

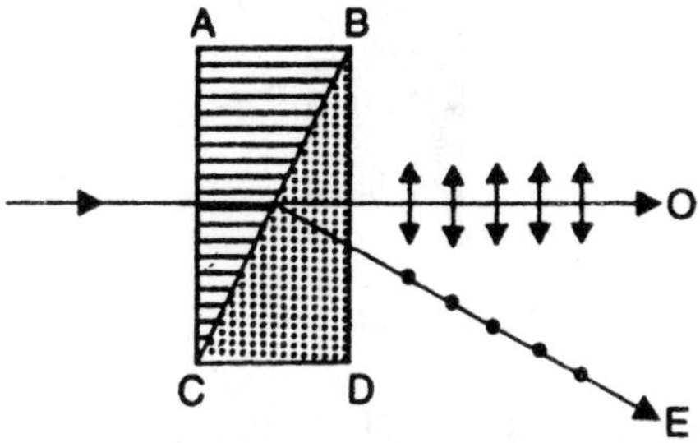

Fig. 3

In the case of calcite, the extraordinary ray will be deviated to the other side. The prisms ABC and BCD are cemented together by glycerine or castor oil. Here, the ordinary emergent beam is achromatic whereas the extraordinary beam is chromatic.

2. Wollaston Prism

It consists of two prisms ABC and BCD of quartz or calcite cut with their optic axes as shown in Fig. 4. They are cemented together by glycerine or castor oil.

A ray of light is incident normally on the face AC of the prism ABC. The ordinary and the extraordinary rays travel along the same direction but with different speeds. After passing BC the ordinary ray behaves as the extraordinary ray and the extraordinary ray behaves as the ordinary ray while passing through the prism BCD. One ray is bent towards the normal while the other is bent away from the normal. In quartz $\mu_e > \mu_o$.

Therefore, the ordinary ray while passing the boundary BC is refracted towards the normal as an extraordinary ray while the extraordinary ray is refracted away from the normal as ordinary ray. If the prisms are made from calcite, the directions of the o-ray and e-ray are interchanged. While coming out of the face BD of the prism, the o-ray and e-ray are diverged.

The prism is useful in determining the percentage of polarisation in a partially polarised beam. Double image prisms are used in spectrophotometers and pyrometers.

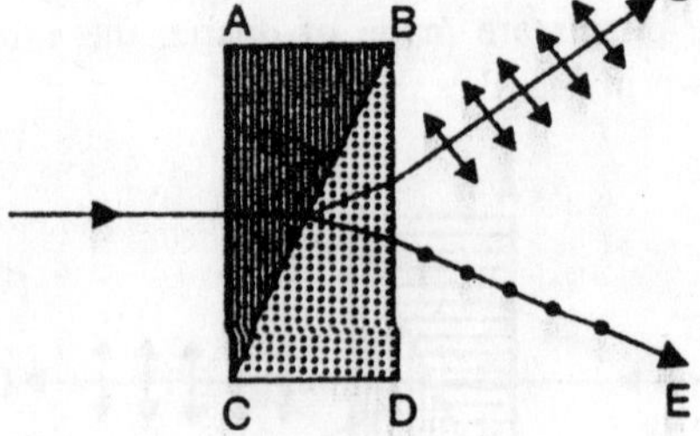

Fig. 4

DEPOSITION POTENTIAL OR DISCHARGE POTENTIAL

Whether a particular ion can be discharged or not, can be found by measuring the separate potential at each electrode during the electrolysis. The potentials of the electrodes are obtained individually by connecting each with a reference, *e.g.*, calomel electrode. The variation of electrode potential with current where D now corresponds to the discharge or deposition potential at the particular electrode.

If an electrolyte contains a number of different ionic species, provided there are no disturbing factors, each ionic discharge will occur as the appropriate potential is reached. Since the process at the cathode involves reduction, cathodic process will occur in the order of decreasing reduction potential.

Let us take the example of electrolysis of NaOH solution.

Anode $OH^- - e^- = OH$

NaOH

cathode $Na^+ + e^- \rightarrow Na + H_2O$

$\rightarrow NaOH + H$

$\downarrow$

$Na^+ + OH^-$

OVER-VOLTAGE

We have already seen that by application of the opposing E.M.F., the cell reaction can be reversed. Usually the opposing voltage required to reverse the cell reaction is much greater than the E.M.F. of the cell and not greater only by an infinitesimal quantity as suggested by the theory. Suppose we have a cell like

$$Pt \mid H_2\ (1\ \text{aim}) \mid HC1\ (a = 1) \mid Pt \mid O_2\ (1\ \text{atm})$$

The voltage of this cell is 1.12 volts. If the opposing voltage is slightly greater than 1.12, the cell reaction should get reversed and evolution of H_2 and O_2 should start. In reality this is not true, and an opposing voltage of 1.7 volts is required. Thus, an excess of 1.7 – 1.12 or 0.58 volt is required, to cause the reverse reaction to the place. *This excess of voltage, over the voltage of the cell that is necessary to cause the reverse reaction, is termed as over-voltage.*

Hydrogen Over-voltage

If a dilute solution is electrolysed with platinised platinum electrodes, hydrogen is liberated at the reversible hydrogen potential of the solution. With electrodes other than platinised electrodes a more potential is required to secure its liberation. This variation in the potential is explained by the difference in polarisation and depends upon the nature of electrodes. *This difference of potential at which hydrogen gas is actually evolved during electrolysis and theoretical value at which it happens is known as hydrogen over- voltage.* At platinised platinum and at zero current density, the hydrogen over-voltage is zero.

Tafel (1905) made measurements with solution freed from oxygen, and found that hydrogen overvoltage at mercury electrodes was a function of current and is given by

$$i = ke^{-\beta v} \qquad ...(1)$$

where i is the current, V the potential of the cathode, k is a constant; characteristic of the electrode and β is given by

$$\beta = \frac{F}{2RT} \qquad ..(2)$$

On taking logarithm of equation (1), we get

$$ln\ i = ln\ k - \beta V$$

$$ln\ i = ln\ k - \frac{VF}{2RT} \qquad \text{[Using eq. (2)]}$$

Differentiating the above equation, we get

$$-\frac{dV}{d\,ln\,i} = \frac{2RT}{F}$$

$$-\frac{dV}{2.303\ d\ \log i} = \frac{2RT}{F}$$

or $$-\frac{dV}{d \log i} = \frac{4.606\ RT}{F} \qquad ...(3)$$

At 18°C, equation (3) becomes as

$$-\frac{dV}{d \log i} = 0.116 \text{ volt} \qquad ...(4)$$

From equation (4), it follows that the cathode potential becomes 0-116 volt more negative for each tenfold increase in the current.

Measurement of Hydrogen Over-voltage

In order to measure hydrogen overvoltage, the solution should be free from oxygen or other reducible material. Then the solution is generally saturated with hydrogen at one atmosphere. Now the current of definite strength is passed through the solution for a sufficient time so as to become constant. The cathode potential is then measured by combining it with reference electrode. *e.g.*, calomel. If hydrogen electrode used in place of calomel electrode, the difference of potential between hydrogen and cathode will give directly the value of over voltage.

Oxygen Over-voltage

At an anode, oxygen is liberated but a more positive value of potential than the calculated value is required to secure its liberation at any appreciable rate. The difference between the two values gives the value of *oxygen overvoltage*. It is generally observed that the metals which possess a high hydrogen overvoltage usually have a low oxygen overvoltage and *vice-versa*. The expression (1) given by Bowden and Tafel holds true for the liberation of oxygen also. The overvoltage values of hydrogen and oxygen are given in the following Table 2.

Table 2

Electrode	Oxygen overvoltage (volt)	Hydrogen overvoltage (volt)
Platinised platinum	*0.00*	*0.25*
Palladium	*0.00*	*0.43*
Gold	*0.02*	*0.53*
Iron	*0.08*	*0.25*
Silver	*0.15*	*0.41*
Nickel	*0.21*	*0.06*

FACTORS AFFECTING OVERVOLTAGE

There are a number of factors which affect the phenomenon of overvoltage:

1. *Current Density* : The influence of current density on over-voltage was predicted by Bowden and Rideal (1928). If hydrogen is not removed by reaction with oxygen or other oxidising agents, the overvoltage w increases with increasing current density I according to the following equation

$$w = a + b \log I, \qquad ...(1)$$

where, a and b are constants. This relationship has been found to be applicable to a large number of cathodes at current densities varying from 10^{-7} to 10^{-1} ampere per sq. cm. It over- voltage is plotted against log I. The results for a number of cathodes in dilute sulphuric acid are plotted. These are in harmony with expectations. The close observation of the curves indicates that the lines in the figure are approximately parallel. The value of b may be represented by

$$b = 2 \times \frac{2.303\ RT}{F} \qquad ...(2)$$

The value of b is equal to 0-118 at ordinary temperatures. Its value as calculated for a number of cathodes like mercury, nickel, aluminium, copper and gold is found to be 0.12.

2. *Hydrogen Ion Concentration (pH)* : From an examination of a variety of experimental data, it is seen that in the absence of strongly adsorbed ions overvoltage is independent of hydrogen ion concentration over a large range of pH values. In the case of strongly acidic or alkaline solutions some deviations occur. These may be due to the large concentration of the hydrogen of hydroxyl ions. It has been found that all substances which alter electro-kinetic potential at cathode surface have an influence on overvoltage.

3. *Temperature* : As we know that over-voltage is due to a slow stage in the ionic discharge process, it follows that overvoltage will decrease with increasing temperature. Generally it has been observed that the change of 2 m.v. per degree is essential to express over voltage in terms of activation energy. If A H?& is the energy of activation of the slow process and $\Delta H\pi_1$ is the rate

constant, the current density I is related to the energy of activation by the following relation

$$I = Br^{-\Delta H\neq/RT}\ e^{\alpha wF/RT} \quad ...(3)$$

where B is a constant and the value of a in eq. (3) may be related to the slope of b in eq. (1). On taking logarithm of the eq. (3), we get

$$ln\ I = ln\ B + \frac{\alpha Fw - \Delta H\neq}{RT} \quad ...(4)$$

Differentiating the above equation with temperature at constant overvoltage, we get

$$\left(\frac{\partial ln\ I}{\partial T}\right)_w = \frac{\Delta H\neq - \alpha\ Fw}{RT^2} \quad ...(5)$$

Also differentiating eq. (4) with respect to temperature at constant current density, I, we get

$$\left(\frac{\partial w}{\partial T}\right)_I = \frac{\alpha Fw - \Delta H\neq}{\alpha RT} \quad ...(6)$$

If the values of a and w are known, it is possible to determine DH ≠ from equation (3) or (4). It is also possible to calculate the value of I_0 *i.e.*, the magnitude of current which flows in each direction at reversible electrode. It is done by plotting w against log I, a straight line is obtained which on extrapolating it to zero over-voltage gives the value of I_0.

4. *Impurities* : Over-voltage is influenced by the presence of various impurities in the cathode materials.
5. *Pressure* : Measurement of hydrogen over-voltage at various pressures have shown that at low pressures over-voltage increases rapidly on the cathodes like copper, mercury or nickel but at high pressures the over voltage is slightly affected.

Theories of Over-voltage

Various theories have been proposed to explain the nature of over-voltage. But it is difficult to give an adequate explanation because different mechanisms may be involved for different metals and at di.ferent current densities. In general over-voltage may be due to one or several slow stages in the discharge of ions at the cathode. These stages are :

(1) Transfer of H^+ ions from the bulk of the solution of the electrode layer.

(2) Transfer of H^+ ions or protons to the electrode.

(3) Neutralization of the charge on the ions or protons by electrons.

(4) Combinations of the resulting hydrogen atoms to form molecules.

(5) Evolution of hydrogen molecules as bubbles of gas.

The theories of over-voltage are based on the assumption that either of the steps (2), (3) or (4) is a slow rate determining step. It is less probable that step (1) plays a part because the over-voltage depends upon the nature of metallic cathode. Let us consider the various theories concerning the various stages from (2) to (5).

(a) *Glasstone, Eyring and Laidler's Theory* : A theory of over-voltage, which regards the slow stage responsible for the hydrogen evolution as the transfer for a proton from the solution to electrode, has been proposed by *Eyring, Glasstone* and *Laidler*. They used the theory of absolute reaction rates. They postulated the following assumptions:

(i) The activation energy for over-valtage is about 10 to 20 Kcal. This is of the same order as that of chemical reaction.

(ii) In the absence of appreciable amounts of neutral salts or when the concentration of hydrogen or hydroxyl ions is not very large, it is found that the over-voltage is independent of pH.

(iii) They deduced that the rate of cathode process is given by the equation of the form

$$I = Fk_1 \ c \ e^{\alpha Fw/RT} \qquad ...(7)$$

where c is the concentration of the species in solution from which protons originate. It is assumed that the value of c remains same in all solutions irrespective of the hydrogen ion concentration, equation (7) may be represented as

$$w = \text{const.} + \frac{RT}{\alpha F} \ln I \qquad ...(8)$$

where w is independent of the pH of the electrolyte The important point of equation (8) is that the transfer of proton takes place only in over voltage portion. The over voltage is operating only in the double layer at the electrode surface. On comparing equations (8) with (3), we observe the following facts:

(i) The quantity Fk_1c in equation (3) is equivalent to B $e^{-\Delta H\neq/R}$ and hence to I_0 *i.e.*

$$Be^{-\Delta H\neq/RT} = Fk_1c = I_0$$

(ii) It also follows that the factor B is equivalent of Fc $e^{-\Delta s\neq/RT}$ multiplied by a proportionality factor.

$$\text{B-proportionality constant} \times Fc\ e^{-\Delta s\neq/R} \qquad ...(9)$$

According to theory of absolute reaction rates, the value of proportionality constant is $\frac{kT}{h}$, *i.e.*,

$$B = \frac{kT}{h} \times Fce^{-\Delta s\neq/R} \qquad ...(10)$$

where k is the Boltzmann constant and h is the Planck's constant. When necessary calculations were made. It was observed that the value of B, *i.e.*, Fc $e^{-\Delta s\neq/R}$ was approximately constant for all the electrodes and had the same value in acid and alkaline solutions. This interesting result reveals that it is the same fundamental process which is responsible for over- voltage at all the electrodes. As the value of B is found to be constant for all the electrodes, it follows from equations (3) and (8) that :

(i) If the metal-hydrogen (M–H) bond is very strong, the activation energy for the transference of electrons will be very small. This is clear that at a platinised platinum electrode the combination of hydrogen atoms should be rate determining process.

(ii) If the metal-hydrogen bond is very weak, it is expected that the values of activation energy and hence over-voltage will be very high. Experimentally, it has been found to be so. Examples are mercury, tin and lead.

(iii) If the cathode metal is one which can adsorb hydrogen readily as in platinum, palladium, nickel or copper, it will form relatively strong metal hydrogen bonds. Therefore, the energy of activation of proton transfer process responsible for overvoltage will be low.

(b) *Neutralisation of Charge or Ion Discharge :* The stage (3), which is the neutralization of hydrogen ion by an electron, has been treated by Gurney and Fowler.

According to classical mechanics, the electron cannot cross the barrier to discharge the ion unless it acquires sufficient energy to pass over the top. However, quantum mechanics (Gurney and Fowler) allows a definite probability of leakage through the barrier of an electron from the cathode to an unoccupied level of the same energy in the ion to be discharged. The current strength which is able to pass at a given potential is then determined by integrating the probability of the transition from any level in the cathode to one of equal energy level in the ion. The relationship for the dependence of current on the over-voltage is of the following form

$$w = a + b \log I$$

where $$b = \frac{3.303\ RT}{2F}$$

In the above equations the value of a should lie between zero and unity and then only the value of b should be 0.5.

The theory due to Gurney and Fowler has been found to be in good agreement with experiments for a number of metals. However, this theory has two weaknesses:

(i) This theory does not give any relation between over-voltage at a particular cathode and the known physical or chemical properties of the solution.

(ii) This theory suffers from the weakness of requiring the combination of an electron with an ion to be a slow process necessitating an activation energy of the order of 10 K cal or more. Actually, it is not so.

(c) *Combination of Atoms to Form Molecules* : According to *Tafel*, the formation of hydrogen molecules from hydrogen atoms is a rate determining step and the over-voltage may be due to the aggregation of hydrogen atoms at cathode. The variation of over-voltage from one metal to another was ascribed to their differing catalytic effects on the rate of combination of hydrogen atoms. Thus, the low over voltage metals such as a platinum, palladium and nickel should be good catalysts and those of high over-voltage such as lead and mercury may be expected to be poor catalysts for the reaction.

$$2H \rightarrow H_2$$

The catalytic effect of the metals is in the following order:

$$Pt < Pd < Fe < Ag < Cu < Pb < Hg$$

Direct experimental studies with atomic hydrogen support this theory.

Objection : This theory due to Tafel does not give the value of b in equation (1). This can be seen from the following treatment.

Suppose the electrode surface is covered with atomic hydrogen. Then the rate of the reaction, $2H \rightarrow H_2$ will be proportional to the square of hydrogen atom concentration, *i.e.*, kn^2 where n denotes the number of adsorbed H atoms per sq. cm. The potential E of the atomic hydrogen electrode may be given as

$$E = \frac{RT}{F} \ln n + \text{constant}$$

Similarly the potential E_0 of molecular hydrogen is given by

$$E_0 = \frac{RT}{F} \ln n_0 + \text{constant}$$

where n, denotes the number of hydrogen atoms which are in equilibrium with normal hydrogen gas at 1 atmospheric pressure. Then the over-voltage w may be put as

$$w = E - Eo$$

or
$$w = \frac{RT}{F} \ln n - \frac{RT}{F} \ln n_0$$

or
$$w = \frac{RT}{F} \ln \frac{n}{n_0} \qquad \text{...(11)}$$

From the above equation it follows that Tafel's theory of over-voltage is independent of hydrogen ion concentration. As $I = kn^2$, equation (11) reduces to

$$w = \frac{RT}{2F} \ln I + a$$

where a is constant. If w is plotted against In I, it gives the slope, b = 2.303 RT/2F, *i.e.* 0.029 at ordinary temperature. But the experimental value of b was found to be 0.12 for various metals. Thus, the Tafel's theory is not in agreement with the experimental value of b.

(d) *Evolution of the Hydrogen Gas :* According to *Mac Innes* and *Alder*, the gas-bubble formation is the essential factor in the over-voltage. They also reported that the potential of a platinised platinum cathode fluctuated during the formation and liberation of a bubble of hydrogen gas. This phenomenon is actually due to super saturation effect and it does not represent any variation of the true over-voltage. However, it is found that the presence of substances, which affects the interfacial tensions, alters the bubble over-voltage. But this is partly due to the fact that the effective current density at the instant of bubble formation is also changed. Thus, the gas bubble formation of an electrode plays a secondary part in the over-voltage process.

Importance of Over-voltage : The existence of over-voltage explains numerous phenomena. Some of these are:

(i) Due to the high over-voltage of lead, it is only deposited on the cathode instead of the hydrogen being evolved during charging in a lead accumulator. This view is confirmed by the fact that if lead is covered with a thin layer of metal having a low overvoltage (say, platinum) and electrolysis is carried out, no lead is deposited but only hydrogen gas is evolved.

(ii) The concept of over-voltage is of great importance in the electrolytic reduction of organic compounds. In the reduction of nitrobenzene, for example, the electrodes with high over-voltage, *i.e.*, lead are used instead of costly platinum electrodes.

(iii) It is also utilised in the industrial production of chlorine and sodium hydroxide by the electrolysis of sodium chloride solution.

(iv) The presence of hydrogen over-voltage makes it possible to deposit metals (zinc, cadmium etc.) electrolytically having more negative potentials than hydrogen from an acid solution.

(v) In the electrolysis of acidulated water, there occurs the preferential liberation of ${}_1H^1$ as compared to ${}_1H^2$ on the platinum electrode. This can be explained on the basis of higher overvoltage of ${}_1H^2$ at the platinum electrode.

POTENTIOMETRIC TITRATIONS

The measurement of the change in potential of an indicator electrode in locating the end points in titrations is well known. The potentiometric method, is especially versatile, because indicator electrodes suitable for the study of almost every reaction employed in titrimetry are now available.

In potentiometric titrations, the change in the electrode potential upon the addition of the titrant are noted against the volume of the titrant added. At the end point the rate of change of potential is maximum. The end point is determined by plotting a curve of potential versus the volume of titrant and recording the inflexion point.

Potentiometric titrations are particularly popular, because they not only establish the equivalence points of many reactions but also yield information about the concentrations of one or more of the reactants during the titrations. Furthermore, this method also provides the possibility for a reasonably complete understanding of a whole reaction. Such information is valuable when new reactants or unusual samples are under investigation. Such titrations have found much favour among analytical chemists due to various noted advantages over other indicator methods.

Advantages of Potentiometric Titrations

1. The necessary apparatus required is generally inexpensive, reliable and readily available in most laboratories.
2. It is easy to interpret titration curves with a minimum of mathematical effort.
3. The method can be used for coloured solutions and in solutions where precipitation occurs during the reaction.
4. The method is applicable for analysis of dilute solutions with high degree of accuracy.
5. One of the great advantages of potentiometric titrations is that several components may be titrated in the same solution without the possibility of indicators interfering with each other, for instance, iodide and bromide may be titrated together.
6. The method may be made automatic by bringing relay in operation which will stop the liquid from running from the burette when E.M.F. reaches a certain value.

Indicator Electrodes : The indicator electrodes most often employed are:

(a) Glass electrode,

(b) Membrane electrode,

(c) Quinhydrone electrode,

(d) Tungsten and molybdenum electrodes,

(e) Hydrogen electrode, and

(f) Antimony-antimony oxide electrode

Types of Titrations

1. *Acid-base Titrations :* Since the neutralization of acids and bases are always accompanied by change in the concentration of H^+ or OH^- ion (*i.e.* pit of the solution), it is evident that hydrogen electrode may be employed in these titrations. The reference electrode used in this case is N-calomel electrode.

A known volume of the acid to be titrated is kept in a beaker having an automatic stirrer. It also has a standard hydrogen electrode. It is connected to a N-calomel electrode through a salt bridge. The hydrogen and calomel electrodes are connected to a potentiometer, and the E.M.F. of the solution is recorded.

After the each addition of base from burette into the beaker, the E.M.F. is measured. This procedure is continued. Then the values of E.M.F. are plotted against the ml. of base added and a curve is obtained.

The potential of any hydrogen electrode is given by

$$E = E_0 - 0.0591 \log OH^+ \text{ at } 25°C$$

where E° is the standard-electrode potential. Also,

$$pH = -\log OH^+$$

$$\therefore \quad E = E^\circ + 0.0591 \text{ pH}$$

Thus, the change in electrode potential or E.M.F. of a cell (made up of the hydrogen and a suitable electrode) is proportional to the change in pH during reaction. The point where the E.M.F. increases at once gives the end-point. This method is not very accurate.

2. *Complexometric Titrations :* Complexometric titrations can be followed with an electrode of the metal whose ion is involved in complex formation.

   For instance, a silver electrode may be used to follow the titration of cyanide ion with a standard silver solution. The potential of the silver electrode may be given by eq. (1) The silver ion concentration will be governed by the equilibrium constant for the complex formation.

$$Ag + 2CN^- \rightleftharpoons Ag(CN)_2^- \quad ...(1)$$

$$K_a = \frac{[Ag^+][CN^-]^2}{[Ag(CN_2^-]} \quad ...(2)$$

The titration curve may be exactly calculated from equations (1) and (2) together with the known concentrations of cyanide and silver. At the equivalence point,

$$2[Ag^+] = [CN^-],$$

So that eq. (3) becomes

$$\left[Ag^+\right] = \sqrt{\left(\frac{K_a}{4}[Ag(CN)_2^-]\right)} \quad ...(3)$$

In many complexometric reaction, the situation cannot be handled so easily because more than one complex is formed, as occurs in the reaction of mercuric ion and cyanide:

$$Hg^2 + 3CN^- = HgCCN)_3^-$$

$$Hg^{2+} + 4CN^- = Hg(CN)_4^{2-}$$

Since 1945, complexometric titrations have become widely used because of the discovery of the metal chelating agents such as EDTA. Potentiometric end-point detection has been successfully applied to EDTA titrations, particularly by Reilley (1956).

3. *Oxidation Reduction Titrations* : Oxidation-reduction (Redox) titrations, unlike the cases discussed above involve the transfer of electrons from the substance being oxidised to the substance being reduced ; for instance,

$$Ce^{4+} + Fe^{2+} = Fe^{3+} + Ce^{3+}$$

If is generally considered that such a reaction consists essentially of two *half reactions* whose standard potentials may be used to calculate the standard potential of the reaction (1)

$$Fe^{2+} = Fe^{3+} + e, E° = -0.76 \text{ V} \quad ...(4)$$

$$Ce^{4+} + e = Ce^{3+}, E° = +1.61\text{V} \quad ...(5)$$

$$Ce^{4+} + Fe^{2+} = Fe^{3+} + Ce^{3+}, E° = +0.85 \text{ V} \quad ...(6)$$

The equilibrium constant (K) of any reaction may be calculated from the following formula:

$$E = \frac{0.0591}{n} \log_{10} K,$$

where n is best defined as the number of equivalents of electricity associated with one molar unit of reaction.

If an acidic ferrous solution is titrated with a standard eerie solution at 25°C, the potential (relative to the N.H.E.) of a platinum electrode in contact with the solution will be given by either of the following equations:

$$E = E^{\circ}_{Ce^{4+}/Ce^{3+}} - \frac{0.0591}{1} \log_{10} \frac{[Ce^{3+}]}{[Ce^{4+}]} \qquad ...(7)$$

$$E = E^{\circ}_{Fe^{3+}/Fe^{2+}} - \frac{0.0591}{1} \log_{10} \frac{[Fe^{2+}]}{[Fe^{3+}]} \qquad ...(8)$$

It would be more convenient to use equation (8) before the equivalence point, as the right hand term of this equation could be easily found from the known extent of the titration. Equation (6) could also be used, but the $[Ce^{3+}]/[Ce^{4+}]$ ratio would have to be calculated by means of the equilibrium constant. Beyond the equivalent point, calculations are most easily made by means of equation (4).

From equation (9) it is evident that the potential at the start of the titration should be $-\infty$ because Fe^{2+} is the only ion present and there are no Fe^{3+} ious— At the mid-point of the titration where $[Fe^{2+}] = [Fe^{3+}]$, equation (9) becomes:

$$E = E^{\circ}_{Fe^{3+}/Fe^{2+}} \qquad ...(9)$$

At the equivalent point, the concentration of unchanged ferrous ion will be equal to the concentration of the unchanged eerie ion. Likewise, the concentration of cerous ion will be equal to the concentration of ferric ion.

Therefore, it is true that

$$\frac{[Fe^{2+}]}{[Fe^{3+}]} = \frac{[Ce^{3+}]}{[Ce^{4+}]} \qquad ...(10)$$

and as

$$K = \frac{[Fe^{3+}]}{[Fe^{2+}]} = \frac{[Ce^{3+}]}{[Ce^{4+}]} \qquad ...(11)$$

It is evident that, at the equivalence point,

$$\frac{[Fe^{3+}]}{[Fe^{2+}]} = \frac{[Ce^{3+}]}{[Ce^{4+}]} = \sqrt{K} \qquad ...(12)$$

Combining equations (6) and (9) with (12), it is found that

$$E_{ep} = E^{\circ}_{Fe^{2+}/Fe^{3+}} + \frac{0.0591}{2} \log_{10} K \qquad ...(13)$$

$$E_{ep} = E^{\circ}_{Ce^{4+}/Ce^{3+}} - \frac{0.0591}{2} \log_{10} K \qquad ...(14)$$

By adding Eqs. (13) and (14). The following expression is obtained for the potential at the equivalence point.

$$E_{ep} = \frac{E^{\circ}_{Fe^{3+}/Fe^{2+}} + E^{\circ}_{Ce^{4+}/Ce^{3+}}}{2}$$

where E_{ep} = end-point.

4. *Precipitation Titrations* : A precipitation titration that involves insoluble salts of metals such as mercury, silver, lead and copper may be followed potentiometrically. The indicator electrode may be made of the metal involved in the reaction or may be an electrode whose potential is governed by the concentration of the anion being precipitated.

The magnitude of the potential change at the end point depends on the solubility of the substance being precipitated as well as on the concentration involved. As an example, the titration of chloride ion with a standard solution of silver nitrate using a silver metal indicator electrode may be considered.

The other electrode used to complete the cell is unimportant provided that it is a true reference electrode, *i.e.*, that it maintains a constant potential In this discussion, it will be assumed that the normal hydrogen electrode (N.H.E.) is used although it is rarely employed in practice and the assumption is convenient because standard potential may be used directly. The potential of the silver electrode will be governed by the appropriate Nernst equation:

$$E_{Ag^+/Ag} = E^{\circ}_{Ag^+/Ag} + \frac{0.091}{1} [\log_{10} (Ag)] \qquad ...(15)$$

As soon as enough silver nitrate solution to precipitate some AgCl has been added, the following equilibrium is established,

$$AgCl \rightleftharpoons Ag^+ + Cl^-$$

the equilibrium constant of which is

$$K_{AgCl} = [Ag^+][Cl^-] = 10^{-10} \qquad ...(16)$$

If 0.1 N sodium chloride is titrated with 0.1N silver nitrate, the silver ion concentration may be considered to be 10^{-9}N as soon as the first few drops of silver nitrate have been added. Equation (15) can be used to calculate the indicator electrode potential.

$$E_{Ag^+/Ag} = 0.80 + 0.059 \log_{10} 10^{-9} = 0.27 \text{ v} \quad ...(17)$$

Likewise, half-way through the titration when the chloride ion concentration has been reduced to 0.033 N.

$$E_{Ag^+/Ag} = 0.80 \text{ V} + 0.059 \log_{10} (3 \times 10^{-9}) = 0.30 \text{ V},$$

At the equivalence point, $[Ag^+] = [Cl^-] = 10^{-5}$N,

$$E_{Ag^+/Ag} = 0.80 \text{ V} + 0.059 \log_{10} 10^{-5} = 0.50 \text{ V}$$

Neglecting any liquid junction potential, the end-point potential (E_{ep}), may also be directly calculated from the following equation,

$$E_{ep} = E^\circ_{Ag^+/Ag} - \frac{0.091}{1} - pK_{Ag} X... \quad ...(18)$$

where pK_{Ag} X is the negative logarithm of the solubility product of any silver halide, Ag X.

Requirement for Successful Potentiometric Titrations : Chemical reactions must fulfil certain specific requirements which are as follows :

(i) The reaction must proceed in one way only, that is any side reaction must be of negligible importance.

(ii) The reaction should be stoichiometric and must proceed almost completely in one direction.

(iii) The rate at which the reaction reaches a state of equilibrium must be involved in performing such a titration.

Variations in Potentiometric Titrations

Pinkhof-Treadwell Method : Pinkhof-Treadwell suggested the use of compensation electrode in place of reference calomel electrode. The potential of such an electrode is exactly equal to the e.m.f. of the indicator electrode at the end point. The end point in such a set-up is located by a sudden reversal of polarity. Hence no potentiometric assembly is needed. A simple *capacity electrometer* can be employed instead of the galvanometer. The mercury in the electrometer moves in one direction when the titration is started and continues till the end point is reached when it stops to move. On further addition of the reagent, the mercury moves in the opposite direction. The *drawback* of this method is that every titrant requires its own electrode system.

Potentiometric titration at constant current : End points in the usual potentiometric titrations are dragged if equilibrium is not established quickly at the indicator electrode. In such cases titrations are carried out successfully at constant current which will result a sharp change of potential at the end point *Kolthoff* (1954) reported the titration of copper (II) using EDTA and a rotating platinum micro electrode at a constant current of 2 micro amperes.

Svoboda (1961) has employed constant current with two platinum electrodes in the titration of a series of organic amines with $HClO_4$. Well defined peaks are obtained at the equivalence point.

Differential titrations : By plotting dE/dV versus P yields peak which gives better result than the step curve obtained by plotting E versus V. The value of dE/dV can be directly obtained by a simple but ingenious device due to *Hall* et. al. If the potential between the two electrodes is now measured which will directly yield the value of dE/dV. After each increment, the nipples is pressed in order to expel the enclosed quantity of liquid and a fresh lot is drawn in from the bulk. The end point will be located by the maximum value of dE. An apparatus for continuous differential potentiometric titrations has been described by Nickolson (1961).

Recent advances : *Meites* and *Goldman* (1983) have modified the equation for acid-base titration curves to incorporate the effect for dilution. They found that the inflexion points always precede the equivalence points in strong acid-base titrations. The relative error would be difficult to locate since it is of the order of 10^{-13} in molar solutions.

POLAROGRAPHY

Introduction : Polarography is an electrochemical technique which makes use of current voltage curves under conditions of concentration polarisation of an indicator electrode. *Polarography is the study of the relationship between the current flowing through a conducting solution and the voltage applied across two electrodes in the solution.* The plot of current against applied voltage for a sample solution is called a *polarogram*.

Heyrovsky devised polarographic method of analysis using dropping mercury electrode. For this work, Heyrovsky was awarded Nobel Prize in 1929. In polarograph, there is a *dropping mercury electrode* which consists of mercury reservoir from which mercury trickles down as small drops through a capillary. This acts as a cathode and is generally called an *indicator or micro electrode.* The anode consists of mercury pool at the bottom of the reservoir and the connection is made by a wire passing through its bottom.

Both cathode and anode are connected across the appropriate ends of a battery. The applied voltage can be varied by adjusting the sliding constant C along a potentiometer wire EF. G is a galvanometer which measures the current strength, S is a shunt for adjusting the sensitivity of the galvanometer G. The reservoir is provided for blowing nitrogen gas through sample. This gas removes dissolved oxygen from the sample. Any dissolved oxygen that remains can be reduced.

As the anode has a large surface area, the polarisation at this electrode is negligible and the potential of the anode may, therefore be constant.

Although various microelectrodes, such as rotating platinum electrode, stationary noble metal electrodes, etc., have been used in polarograph. Yet the dropping mercury electrode is the best for obtaining the current-voltage curves due to the following advantages:

(i) Its surface area is reproducible with any given capillary.

(ii) Mercury possesses the property of forming amalgams with many metals and hence lowers their reduction potentials.

(iii) From the weight of the drops, the surface area can be calculated.

(iv) The diffusion current assumes steady values almost instantaneously and is reproducible.

(v) High over-voltage of hydrogen on mercury makes possible the deposition of ions which are difficult to be reduced.

The dropping mercury electrode is useful over the range +2.4 to –2.8 volt. 1262

The above advantages largely out-weigh a few minor disadvantages, the most important of which are :

(a) The cathodic varsatility of mercury is not matched by its anodic behaviour, mercury dissolution setting in at about 0.4V (saturated calomel electrode). Usually as far metal ions are concerned, this is of little importance, since almost all metal ions reduce at potentials considerably more negative than this.

(b) Oxygen is reduced by a two-stage process which shows a current-voltage curve extending over most of the cathodic working range. Dissolved oxygen must be therefore removed from a solution by flushing with some inert gas such as nitrogen, before electrolysis is attempted.

(c) Mercury is toxic but sensible precautions make its use safe.

There are two types of polarographs :

(i) *Manually operated polarographs*

(ii) *Automatic polarographs which record the current-voltage curve automatically*. The record is made photographically on a sheet of sensitized paper fixed to a drum which can be rotated by a small motor.

Polarographic Cells

Working : Suppose an external E.M.F. is applied to a polarographic cell having a solution of cadmium chloride. All the positively charged ions present in the solution will be attached to the dropping mercury electrode by an *electrical force*, and by a *diffusive force* which results from the concentration gradient formed at the surface of the electrode. Thus, the total current flowing through the cell may be regarded as the sum of electrical force and diffusive force.

Suppose we increase the applied voltage and record the current. This is known as *residual current* which is the current carried by the supporting electrolyte and impurities present in the sample. At point B, the potential of the electrode becomes equal to the decomposition potential of the Cd^{2+} ions. The current suddenly increases along the curve BC. C is a point where current no longer increases with increasing applied voltage but reaches a steady value at a point D. After this no increase in current is observed at higher cathode potentials. Thus, the current corresponding to the curve CD is known as *limiting current*. The difference between the residual current and the limiting current is called *diffusion current*. The diffusion current is generally denoted by i_d.

Factors Affecting the Limiting Current

Limiting current is the sum total of four contributory factors which are as follows:

(i) Residual or condenser current.

(ii) Migration or electric transference current.

(iii) Diffusion current.

(iv) Kinetic current.

Let us discuss these one by one.

(1) *Residual or Condenser Current* : If a potential is applied to a solution of potassium chloride by using a cathode of dropping mercury electrode, the potassium ions will be attracted to the cathode; this layer of positive ions will in turn attract the negative ions and an electrical double layer of cations and anions will be set up around the drop of mercury.

At any potential, the magnitude of the current indicates the energy required to set up the electrical double layer of cations and anions and to maintain this structure against the disordering effects of the growth of the drop and the Brownian motion in the solution. "*The electrical double layer resembles an electrical condenser and the current required to develop and maintain this layer is called condenser current.*"

(2) *Migration Current* : Electroactive material reaches the surface of electrode largely by two processes, viz.

(i) One involves the migration of charged particles in the electrical field caused by the potential difference existing between the electrode surface and the solution.

(ii) The other is concerned with the diffusion on particles.

The current required for the above two processes is known as migration current. Heyrovsky showed that the migration current can be practically eliminated if an indifferent electrolyte is added to the solution in a concentration so large that its ions carry essentially all the current. The indifferent electrolyte is also known as supporting electrolyte. Its concentration should be at least 100-fold that of the electroactive material.

(3) *Diffusion Current* : This current is directly proportional to the concentration of the substance being reduced or oxidised at the dropping electrode. The equation which describes the polarographic diffusion current was first derived in 1934 and is known as the Ilkovic equation. This equation is:

$$i_d = 607\ nD^{1/2}Cm^{2/3}\ t^{1/6}$$

where i_d is the average diffusion current (in microamperes) during the life of a drop, n is the number of faradays of electricity required per mole of the electrode reaction, D is the diffusion coefficient of the oxidisable or reducible substance in units of $cm^{-2}\ sec^{-1}$.

C is the concentration of the oxidisable or reducible substance in millimoles per litre, m is the rate of flow of mercury from the dropping mercury electrode expressed in the units of nag/sec., and t is the drop time in seconds.

"The observed diffusion current is directly proportional to the concentration of electroactive material. This fact is the basis of quantitative polarographic analysis."

Because equation (1) neglects the curvature of the electrode surface it cannot be expected to be highly accurate. Correction for curvature have been applied to the Ilkovic equation by Lingaoe and Loveridge. The modified equation (1) is as follows:

$$I_d = 607\ nD^{1/2}m^{2/3}\ t^{1/6}\ (1 + 39\ D^{1/2}m^{-1/3}\ t^{1/6})$$

The term in bracket is corresponding to the difference between linear and spherical diffusion. This correction term is not large, amounting to about 3–7% of the total current. This equation is used for accurate work.

(4) *Kinetic Current :* The limiting current may be affected by the rate of non-electrode reaction called the kinetic current.

This current results if the oxidised or reduced form of the electroactive species involved is in a chemical equilibrium with other substances. Since these are rate processes, the current resulting from this is called the kinetic current.

The Half-Wave Potential

The half wave potential in polarography is a characteristic property of the electroactive material. *This potential is found on the steeply rising portion of the current voltage curve and is one half of the distance between the residual and limiting currents. It is called the half wave potential,* $E_{1/2}$.

This significance of the half wave potential can be best understood by considering the reduction taking place at the dropping electrode which may be represented as follows:

$$Ox + ne \rightleftharpoons Red$$

The reversible potential of this system may be defined by Nernst equation as

$$E_{d.e.} = E^{o}_{oxired} - \frac{0.059}{n} \log \frac{[red]}{[ox]} \qquad ...(1)$$

In the electrolysis of the reducible ion, the concentration of the oxidised form will start to decrease as soon as the E.M F. becomes large enough to reduce it and the concentration of the oxidised form then starts to decrease. Electroactive ion from the solution will then move into the volume around the electrode which was depleted of these ions, and a concentration gradient builds up between the bulk of the solution and the volume around the electrode. The current on the voltage-current curve may be expressed as:

$$i = k_{ox} ([OX]-[OX]^{\circ}) \qquad ...(2)$$

where i is the average current flowing through the cell, [OX] is the concentration of the reducible ion in the bulk of the solution, $[OX]^{\circ}$ is the concentration of oxidant at the electrode-solution interface, and kyx which includes the capillary characteristics and the n, m and t terms of the Ilkovic equation.

As soon as the current attains the limiting current, the concentration of the oxidised form at electrode solution interface becomes zero, and thus equation (2) becomes as

$$i_d = k_{ox} [OX] \qquad ...(3)$$

On combining equations (2) and (3), we obtain the concentration of [OX] at the surface of the electrode as follows :

$$[OX]^{\circ} = \frac{i_d - 1}{k_{ox}} \qquad ...(4)$$

For metals that form amalgams, the concentration of the metal amalgam is directly proportional to the current on the current voltage curve, *i.e.*.

$$[red]^{\circ} = \frac{i}{k_{red}} \qquad ...(5)$$

Substituting equations (4) and (5) in (1), we get the value of potential of the dropping electrode.

$$E_{d.e.} = E_{ox,\,red} - \frac{0.059}{n} \log \frac{i}{i_d - 1} - \frac{0.059}{n} \log \frac{k_{ox}}{k_{red}} \qquad ...(6)$$

According to the definition of half wave potential, it is that potential at which the current is half of the diffusion current, *i.e.*. $i = i_d/2$; then equation (6) becomes

$$E_{d.e.} = E_{1/2,} = E^o_{ox,\, red} - \frac{0.059}{n} \log \frac{k_{ox}}{k_{red}} \quad ...(7)$$

When equation (7) is substituted back into equation (6), the equation for the potential at the dropping electrode in the reversible polarographic wave is given by

$$E_{d.e.} = E_{1/2,} - \frac{0.059}{n} \log \frac{i}{i_d - i} \quad ...(8)$$

Criterion of Reversibility : The potential of the dropping electrode may be defined by the equation (8) which is

$$E_{d.e.} = E_{1/2,} - \frac{0.059}{n} \log \frac{i}{i_d - i}$$

where $E_{d.e.}$ is the E.M.F. of the dropping electrode. $E_{1/2}$ is the half wave potential, n the valence change in the electrode process, i the current at any point on the polarogram and id the diffusion current in the electrode process.

In order to test the reversibility or irreversibility of an electrode process, the best way is to plot $E_{d.c.}$ vs. log $i/(i_d–I)$. If there is a straight line with a slope equal to –0-059/n volts it is generally safe to conclude that the electrode reaction is reversible. In this method the error of ±10% is allowed. If the deviation is greater than ±10%, it may be assumed that the electrode process is irreversible.

The oscillations are as a result of the successive appearance of new drops at intervals of a few seconds. Each step is called a *polarographic wave* characterised by its half wave potential, $V_{1/2}$ and diffusion current i_d^{max}.

APPLICATIONS OF POLAROGRAPHY

The technique has found several applications because it is a convenient method for measuring the electrode potentials and for studying electrode reactions. Some application are :

(i) Polarography can be used to estimate cations and anions in the presence of various interfering ions. For example, lithium having a negative half-wave potential can be estimated in the presence of other alkali metals, provided the latter metals are not of high concentration.

(ii) One of the important applications of the polarography is the direct determination of dissolved oxygen in solution, aqueous or organic. In this procedure, of course, the dissolved oxygen is not swept out with nitrogen gas as with other samples. Instead the oxygen waves (there are two) are measured and the dissolved oxygen determined.

The method suffers on interference from combined oxygen in the sample. For example, in aqueous solution, the oxygen combined in the H_2O molecule does not take part in a reaction. This is a distinct advantage over other methods for determining oxygen, such as neutron activation analysis, which measures the total amount of oxygen present.

(iii) Another application is the determination of oxidation states of a metal ion in solution. For example, in a mixture of ferrous ion and ferric ion (Fe^{2+} and Fe^{3+}) each ion can be determined quantitatively. This is a direct advantage over most methods of elemental analysis.

(iv) *Determination of the instability constants of complexes.* Polarography is a powerful tool in the study of the composition of complexes if the simple metal ion and complex of that metal ion in the same oxidation state give reversible processes and involve the same number of electrons. We shall consider the reduction of a complex ion to the ligand ion and the metal as follows:

(i) For the simple metal ion, the reduction is as follows:

$$M^{n+} + ne + Hg \rightarrow M\,(Hg) \qquad ...(1)$$

For the potential of the metal amalgam electrode in a solution having simple M^{n+} ion, we can write

$$E_{d.e.} = (E_{1/2})_m - \frac{0.059}{n}\log\frac{[M(Hg)]}{[M^{n+}]} \qquad ...(2)$$

where M(Hg) is the concentration of the metal, and M, in the amalgam.

(ii) In the case of complex ion of the same metal, M, the reduction is

$$MX_p^{+(n-pb)} + ne + Hg \rightarrow M(Hg) + pX^{-b} \qquad ...(3)$$

The potential of the metal amalgam electrode can be written as

$$E_{d.e.} = (E_{1/2})_c - \frac{0.059}{n} \log \frac{[M(Hg)][X^{-b}]^p}{[MX^{+(n-bp)}]} \qquad ...(4)$$

On multiplying the right-hand side of equation (5) by $[M^{n+}]/[M^{n+}]$, we get

$$E_{d.e.} = (E_{1/2})_c - \frac{0.059}{n} \log \frac{[M(Hg)][X^{n+}][X^{+b}]^p}{[M^{n+}][MX_p{}^{+(n+pb)}]} \qquad ...(6)$$

Equation (2) can be written as

$$E_{d.e.} = (E_{1/2})_c - \frac{0.059}{n} \log \frac{[M(Hg)]}{[M^{n+}]} \qquad ...(7)$$

Substituting this value in equation (6), we get

$$E_{d.e.} = (E_{1/2})_c - \left[E_{d.e.} - (E_{1/2})_m\right] - \frac{0.059}{n} \log K \qquad ...(8)$$

where K is the instability constant of the complex

$$MX_p{}^{+(n+Pb)} \rightarrow M^{+n} + pX^{-b}$$

and its value is given by this equation as

$$K = \frac{\left[M^{+n}\right]\left[pX^{-b}\right]^p}{\left[MXp^{+(n-pb)}\right]} \qquad ...(9)$$

Equation (8) can be rearranged to give

$$\frac{0.059}{n} \log K = (E_{1/2})_c - E(_{1/2})_m$$

From equation (9), the value of instability constant K can be calculated if the half wave potentials of metal complex $(E_{1/2})_c$ and metal $(E_{1/2})_m$ are known.

VOLTAMETRY

As the name implies, both current and potential are measured in voltametry. The voltametric experiment involves an indicator electrode and a reference electrode. A potential difference is imposed between these electrodes, and the current that flows because of electrochemical reactions is measured. The output consists of a record of current as a function of indicator electrode potential. In all these experiments one deals with heterogeneous reactions that occur at the surface of the working or indicator electrode. Since material must get from the bulk of solution to the electrode in order to react, a primary consideration in voltametry is mass transport in solution.

The transport of charged particles in solution can be explained on the basis of migration, convection and diffusion. The fact that diffusion and concentration are intimately connected is the basis of analytical utility of voltametry. The experiments are carried out in such a way that only diffusion makes an important contribution to the movement of material to the electrode. It is necessary to mention here that the term polarography is applied to a voltametric experiment carried out using a dropping mercury electrode as the working electrode.

AMPEROMETRIC TITRATIONS

In amperometric titrations, the voltage applied across the indicator electrode and reference electrode is kept constant and the diffusion current passing through the cell is measured and plotted against the volume of reagent added. The diffusion current is generally proportional to the concentration, hence the titration curve consists of two straight lines. The point of intersection of the two straight lines indicates the end point. Amperometric titrations are also known as *polarographic* or *polarometric titrations.*

Description of the Apparatus

Amperometric titrations may be performed either with the dropping mercury electrode, or with rotating platinum electrode. For dropping mercury electrode, the conventional polarographic equipment is used with cells modified to allow the addition of titrating reagent. Calomel electrode is used as a reference electrode. There is a galvanometer to measure the current. A series rheostat may be used for varying the sensitivity of the galvanometer. In order to carry out the amperometric titration, the voltage applied to the indicator electrode and the reference electrode is kept constant and a sensitive galvanometer is made to indicate the value of diffusion current after each increment of the titrant has been added. In place of dropping mercury electrode a platinum rotating micro-electrode is used. The latter electrode possesses certain advantages over the dropping electrode which are as follows:

(i) It is simple to construct.

(ii) It increases the workable range on the positive voltage side upto 0.9 V. Thus, it can be used at positive potentials where the mercury anode may not be used.

(iii) The rotation of the electrode increases the value of diffusion current as much as 20 times the value in polarography. Hence the technique becomes more sensitive.

The dropping mercury 'electrode was introduced by Laitinen and Kolthoff (1941). It consists of a glass tube of about 15-20 cm, in length and 6 mm. in diameter. It is usually fabricated with 5-10 mm length of Pt-wire extending horizontally from its vertical glass supporting tube. The electrode is mounted in the shaft of a motor and rotated at a constant speed of 600 rounds per minute. In amperometric titrations, removal of oxygen is necessary if the electrolysis is conducted at an E.M.F. at which oxygen would have a diffusion current. Removal of oxygen is done by bubbling hydrogen or purified nitrogen before the commencement of titration and for 1 minute after each addition of the titrant.

Technique Involved in Amperometric Titrations

The technique of amperometric titration may be illustrated by a specific example, *i.e.*, the titration of a reducible ion, such as Pb^{2+} with a non-reducible ion, such as SO_4^{2-} Lead ions being reducible at the cathode gives a diffusion current whereas the sulphate ion being non-reducible shows no diffusion current. If the voltage is maintained constant at any value of diffusion current plateau, the limiting current will be represent- ed by i_0 corresponding to 'C_0' the initial concentration of lead ions. The titrant exhibits no diffusion current at the applied E.M.F. Incremental addition of SO_4^{-2} removes some of the electroactive Pb^{2+} ions.

Advantages of Amperometric Titrations

These titrations have a number of advantages over other methods, say potentiometric methods, polaragraphic methods and other electroanalytical methods. Some of these are as follows :

(i) The equipment used is very simple. The characteristics of electrodes are less important.

(ii) Since the method is relative, fewer disturbing factors are prevalent.

(iii) Accuracy is higher than in polarography, and the error in end point is determined by one's ability to measure the volume of titrant

(iv) The method has a greater sensitivity than such methods as potentiometric and conductometric processes and concentrations of 0.1 to 0.00001 M are handled easily.

(v) The method has general adaptability to precipitation reactions.

(vi) It is not necessary to determine the capillary characteristics of dropping electrode. The use of rotating electrode is possible.

(vii) It is not essential to maintain constant current throughout the titration.

(viii) It is not necessary that a reaction occurring during the titration be reversible.

(ix) The depolarising substances cannot be estimated very accurately by polarography. However, these can be estimated accurately by amperometric titrations. For such substances, a suitable titration is selected which gives a diffusion current.

The substances, which do not possess equilibrium potentials, may be studied by amperometric methods. An interesting example is that magnesium does not give the diffusion current curve but it can then be estimated by 8-hydroxyquinoline which possesses the reduction curve.

(x) The amperometric titration can usually be carried out rapidly because the end point is found graphically.

(xi) In these titrations, foreign salts do not cause any interference. Generally these are added as the supporting electrolyte in order to eliminate the migration current.

(xii) Although, a polarograph is generally used to apply the voltage to the cell, its use is not essential in amperometric titrations. The constant applied voltage may be obtained with a simple potentiometric device.

(xiii) The titrations are also possible to carry out if the substance formed in the reaction is slightly soluble as in precipitation titration or hydrolysed as in the case of acid-base titrations. The reason for this is that the readings near equivalence point have no special significance in amperometric titrations. Readings are recorded in regions where there is excess is suppressed by the mass-action effect; the point of intersection of these lines will give the equivalence point.

Disadvantages of Amperometric Titrations

Inspite of many advantages as discussed above, amperometric titrations possess some disadvantages which may be discussed as follows:

(i) Due to the effect of co-precipitation and post-precipitation the amperometric titrations sometimes give inaccurate results.

(ii) Due to the presence of foreign substances, the relative change of current during the titration is very little in amperometric titrations.

FUEL CELLS

Recent advances in electro-chemistry and technology have led to the introduction of compact source of power in which fuels are used to produce electricity without the intervention of thermal devices such as generators, turbines, etc. These fuel cells are much more efficient than thermal sources. These are used in space-craft and may become the basis of pollution-free transport.

In fuel cells an attempt is made to make the fullest possible use of the free energy of reactions, such as the combustion of fuels, to produce electrical energy. Processes are chosen which occur as nearly reversible as possible in order to obtain the maximum useful proportion of ΔG.

The mode of operation of fuel cells is fundamentally different from that of batteries. While batteries store electrical energy, fuel cells convert energy obtainable from chemical process directly into electricity,

The ideal efficiency, e of a fuel cell is given by

$$\in = \frac{\Delta G}{\Delta H} = \frac{E}{E - T\left(\frac{\partial E}{\partial T}\right)_p} \quad ...(1)$$

where ΔH being the heat of the reaction used. This equation makes it clear that for a cell with a positive temperature coefficient supply of external heating will, in principle, provide an efficiency greater than 100%.

Fuel cells are electrochemical devices-which convert the energy of fuel oxidation reactions into electrical energy. A fundamental, but important, example of a fuel cell is the hydrogen-oxygen cell. Fuel cells based on combustion of hydrocarbons are also operated. These are discussed below:

(i) Hydrogen-oxygen Cell

It consists of two electrodes made of porous graphite. Platinum is coated on the surface of the electrodes. The electrodes are placed in aqueous solution of KOH or NaOH. Hydrogen and oxygen are bubbled into the cell under a pressure of about 50 atm. When the electrodes are connected. a flow of electric current takes place. The electrode reactions are :

Anodic Reaction

$$H_2 \rightarrow 2H^+ + 2e$$

$$2H^+ + 2OH^- \rightarrow 2H_2O$$

The net half cell reaction is

$$H_2 + 2OH^- \rightarrow 2H_2O + 2e^-$$

Cathodic reaction

$$1/2O_2 + H_2O + 2e^- \rightarrow 2OH^-$$

Now, the overall cell reaction is

$$2H_2(g) + O_2(g) \rightarrow 2H_2O$$

The e.m.f. of the cell is found to be 1.23 volt. The cell reaction is the same as combustion of H, in air or oxygen. The energy of the fuel oxidation reaction has not been liberated as heat but it has been directly converted into electrical energy.

For an efficient cell all processes must occur rapidly. Reactants must be able to reach the electrodes easily so that porous electrodes with large internal surface area, saturated with electrolyte, are used. The pore sizes are often graded from large on the gas side of an electrode to small on the electrolyte side. Good catalyst materials ensure rapid electrochemical reactions and, in combination with a higher temperature, help to suppress the cathodic formation of perhydroxyl ions by the reaction

$$O_2 + H_2O + 2e \rightarrow OH^- + HO_2^- \qquad ...(2)$$

With such precautions the hydrogen-oxygen cell can be made to show efficiencies of upto 75%. A Bacon-type cell has been successfully employed in space projects where the water produced at the rate of about a pint per kilowatt hour-is used to supplement the water supply.

Many other fuel cell systems have been devised with varying degrees of success. Provided that certain inherent difficulties can be overcome, a wide variety of designs and modes of operation could become available for specific purposes. Their use for traction purposes replace engines with high pollution rise is a major attraction. At the other extreme artificial hears, powered by fuel cell consuming food fuels, have been suggested as a further possibility.

At present, a great drawback is that while the attractive prospect of the use of cheap fuels such as hydrocarbons presents itself, difficulties are encountered in practice by the poisoning of catalytic surfaces by intermediates.

(ii) Hydrocarbon-oxygen Fuel Cells

Hydrocarbon and their oxygenated derivatives such as CH_4, C_2H_6, C_3H_6, alcohols etc. are often used as fuels cells. The half cell reactions with propane are given below:

Anodic reaction

$$C_3H_8 + 6H_2O \rightarrow 3CO_2 + 2OH^- + 20e^-$$

$$H^+ + OH^- \rightarrow H_2O] \times 20$$

Cathodic reaction

$$O_2 + 2H_2O + 4e^- \rightarrow 4OH^-] \times 5$$

The over-all reaction

$$C_3H_8 + 5O_2 \rightarrow 3CO_2 + 4H_2O$$

Hydrocarbon fuel-air fuel cells normally use platinum electrode, concentrated phosphoric acid electrolyte, and operate at temperatures in the region of 373 K. These cells can produce about 0.1 W cm^{-2} of electrode surface.

Thermodynamic characteristics of some possible fuel cells are given in the Table 3.

Table 3 : Thermodynamic characteristics of some fuel cells

Cell	Cell reaction	ΔG^o (kJ mol^{-1})	ΔH^o (kJ mol^{-1})	E (V)	$\in$
H_2–O_2	$H_2+1/2O_2 \rightarrow H_2O$	– 237.2	– 258.9	1.229	0.83
C–O_2	$C+O_2 \rightarrow CO_2$	– 137.3	– 110.5	0.712	1.24
CH_4–O_2	$CH_4+2O_2 \rightarrow CO_2+H_2O$	– 818.0	–890.4	1.060	0.92
CH_3OH–O_2	$CH_3OH+3/2O_2 \rightarrow CO_2+2H_2O$	– 706.9	– 764.0	1.222	0.93

Hydrogen-Oxygen Fuel Cells in Manned Space Flights

The hydrogen-oxygen fuel cells are used in some of the manned space flights. The electrolyte used in these cells is an ion-exchange material and not a solution of KOH or NaOH.

The ion-exchange material which is used in the form of a membrane allows easy passage of protons which react with oxygen and electrons to form water.

There are a number of engineering problems which shall have to be solved before fuel cells become practical sources of electric energy. Once it is done, fuel cell technology would bring revolution in the area of energy production.

WAVE PROPAGATION AND MOMENTUM CONSERVATION

A plane wave propagating through a vacuum in a general direction is represented by the equation (1).

$$E = E_0 e^{i(\omega t - k.r)} \quad ...(1)$$

where k is the vector that defines the direction of propagation of the wave, k can be written as

$$k = ik_x + j\,k_y + k\,k_z \quad ...(2)$$

When a light wave interacts with an isotropic material, it will be affected in the same way whatever may be the direction of the beam with respect to the material. Hence, the refractive index of the material is the same in every direction and the velocity of the beam passing through the material is also the same in every direction. Since the magnitude of k is related to the velocity, the components k_x, k_y and k_z would all be the same in case of isotropic crystals. In anisotropic crystals, the light beam experiences different refractive indices in different directions and propagates at different velocities in different directions. Therefore, the components of k will have different values in different directions.

If we consider the photon description of light beam, we know that the momentum and wavelength of a photon are related through the following expression.

$$p = \frac{h}{\lambda} = \frac{2\pi h}{2\pi\lambda} = \hbar k$$

Momentum p is a vector quantity and hence we write the above equation in vector form as

$$p = \hbar k \qquad ...(3)$$

where k is the vector which defines the propagation direction of the wave associated with the photon. Thus, the significance of k in eqn. (1) and (3) is the same.

When a beam of light passes from air into a crystal, the momentum of the wave must be conserved at the interface of the two media. It requires that the normal component and parallel component of the momentum must be separately conserved at the boundary. Therefore, the transverse component of the momentum must be continuous across the boundary. There cannot be an abrupt jump or change in the transverse momentum of the wave as the beam enters the medium. In view of the relation (3), it means that the transverse component of k should not change abruptly at the boundary.

LINEAR MEDIUM

The optical media are basically dielectric materials. They do not allow electric current to pass through when an electric field is applied across them. Instead, they get polarized. The electric field exerts forces on the valence electrons. These forces are quite small, and induce *electric dipoles* in the medium. These dipoles orient in the direction of the electric field and the dielectric is said to be *polarized.*

Dielectrics are polarized, for example, when they are placed between the charged plates of a capacitor. The polarization vector, P, denotes the extent to which the dielectric is polarized. The electric polarization is parallel with and directly proportional to the applied field, E. It is given by

$$P = \varepsilon_o \chi E \qquad ...(1)$$

where ε_o is the permitivity of free space and χ [$= (\varepsilon/\varepsilon_o) - 1$] is a dimensionless constant known as the *electric susceptibility* of the medium. The stronger the electric field, the greater will be the polarization and a plot of P versus E is a straight line. Materials, in which such kind of linear relationship holds, are known as *linear dielectrics.*

Light waves are electromagnetic waves and when they propagate through a dielectric, the electric field of the waves polarizes the dielectric. Hence the optical parameters of the dielectric are closely related to the dielectric polarization. The refractive index of the medium is given by,

$$\mu = \sqrt{\varepsilon} = \sqrt{1 + 4\pi\chi} \qquad ...(2)$$

As long as the intensity of the light propagating in the dielectric medium is small, the parameters χ and μ are constant quantities and are independent of the intensity of light. Ordinary light sources generate light of field strengths of the order of 10^5 V/m which are very small compared to atomic fields, and therefore cannot affect the optical parameters of the medium.

NONLINEAR POLARIZATION

When high electric field strength is used, it is expected that P cannot increase linearly indefinitely with E and will become saturated. Therefore, we may anticipate nonlinear behaviour of Pat very high field strengths. Lasers produce light of field strengths of the order of 10^7 to 10^{11} V/m, which are of the order of the atomic field strengths. Therefore, the intense light of lasers is in a position to cause nonlinearity of P and influence the optical parameters of the medium. When the electric field E in the light is very large, the parameters χ, ε and μ become the functions of E. Since the directions of P and E coincide in an isotropic medium, we can express χ as a power series in the field strength as,

$$\chi(E) = \chi_1 + \chi_2 E + \chi_3 E^2 + \ldots \quad \ldots(1)$$

and as $$P = \varepsilon_o \chi(E) E$$

$\therefore$ $$P = \varepsilon_o(\chi_1 E + \chi_2 E^2 + \chi_3 E^3 + \ldots) \quad \ldots(2)$$

where χ_1 is linear susceptibility and is much greater than the coefficients of the nonlinear terms χ_2, χ_3 and so on. The nonlinear terms contribute noticeably only at very high-amplitude electric fields.

The second order nonlinear polarization is given by

$$P_2 = \varepsilon_o \chi_2 E^2 \quad \ldots(3)$$

and third order nonlinear polarization by

$$P_3 = \varepsilon_o \chi_3 E^3. \quad \ldots(4)$$

Nonlinear polarization leads to nonlinear optical effects. Materials, in which polarization exhibits nonlinear dependence on the field strengths, are called *nonlinear media*. The nonlinear variation of electric polarization with the electric field strength in a nonlinear medium. In optically isotropic materials, the coefficients of even powers of E in Eqn. (3) are zero. But in case of anisotropic materials, coefficients of both odd and even powers of E exist. Strictly speaking, any medium becomes nonlinear provided the electric field of the incident radiation is very high. With the advent of intense lasers, the *nonlinear* polarization has assumed importance and made it possible the phenomena of frequency conversion and other non-linear effects.

Now if we take into the consideration of electric field oscillations of a light wave, we can write for the electric field of the wave as

$$E = E_o \sin \omega f$$

The polarization caused by this wave can be expressed as

$$P = \varepsilon_o (\chi_1 E_o \sin \omega t + \chi_2 E_o^2 \sin^2 \omega t + \chi_3 E_o^3 \sin^3 \omega t + ...) \quad ...(5)$$

The above equation can be re-written as

$$P = \varepsilon_o [\chi_1 E_o \sin \omega t + 1/2\, \chi_2 e\, E_o^2 (1 - \cos 2\omega t) + 1/4\, \chi_3 E_o^3 (3 \sin \omega t - \sin 3\omega t) + ...] \quad ...(6)$$

Thus, as the electric field in the incident wave oscillates, the array of dipoles produces an electromagnetic wave, which may be called a *polarization wave.* When the incident light beam is not intense, the polarization wave will be in phase and have the same frequency as the incident light wave.

The higher order terms in eqn. (6) indicate that for higher beam intensities, polarization waves of higher frequencies also will be produced. We have to consider the phases of the waves at these other frequencies in order to determine the frequencies that get enhanced.

SOLVED EXAMPLES

Example 1:

The E.M.F. of the cell. Calomel Electrode | HCl of unknown pH | Quinhydrone electrode is 0.25 volt at 26°C. The E.M.F. of the calomel electrode is – 0.24 volt and normal electrode potential of the quinhydrone is – 0.699 volt. Find the pH of the HCl solution.

Solution:

$$E = E^o_{\text{C lomel–Quinhydrone}}$$

$$= E^o\ \text{Calomel} - \left[E^o_{\text{Quinhydrone}} - \frac{2.303\ RT}{E} \log C_{H^+} \right]$$

$$= E^o\ \text{Calomel} - E^o_{\text{Quinhydrone}} - 0.0591\ \text{pH at } 25^oC$$

$$\therefore \qquad pH = \frac{E^o_{Calomel} - E^o_{Quinhydrone}}{0.0591}$$

or
$$pH = \frac{-0.24 + 0.669 - 0.25}{0.0591}$$

$$= \frac{0.179}{0.0591} = 3.028.$$

Example 2:

The E.M.F. of the cell H_2 | N/25 Aniline hydrochloride solution | Hg_2Cl_2 in N – KCl Hg is 0.485 volt at 25°C. Calculate the pH value of the solution and the degree of hydrolysis of N/25 aniline hydrochloride solution at 25°C. The E.M.F. of the calomel electrode = – 0.281 volt.

Solution:

$$E = EH_2 - EHg_2Cl_2$$

$$= \left[E^o_{H^2} - \frac{2.303\ RT}{F}\log C_H^{\ +}\right] - E_{Hg_2}Cl_2$$

$$= 0 - 0.0591 \log C_{H+} + 0.281$$

$$= 0.0591\ pH + 0.281$$

$$\therefore \qquad pH = \frac{E - 0.281}{0.0591} = \frac{0.485 - 0.281}{0.0591} = \frac{0.204}{0.0591} = 3.4517$$

$$\therefore \qquad \frac{1}{\left[H^+\right]} = \text{Anti log } 3.4517 = 2830 \text{ or } H^+ = \frac{1}{2830}$$

In N/25 aniline hydrochloride, concentration of

$$H^+ \text{ or } \quad HCl = \frac{1}{2830}$$

$$\therefore \quad \text{Degree of hydrolysis} = \frac{1}{2830} \times 25 = \frac{1}{113.2}$$

Hence salt is $\frac{100}{113.2}$ = 0.8834 % hydrolysed.

Example 3:

Determine the solubility of AgCl from the following cell :

Ag | AgCl (sat)m KCl (0.1M) || $AgNO_3$ (0.1M | Ag

The E.M.F. of the cell was found to be 0.4550 volt at 25°C. The mean activity coefficient of 0.1 M $AgNO_3$ is 0.82 and that of 0.1 M KCl is 0.76.

Solution:

The E.M.F. of the above cell is given by

$$0.4550 = \frac{2.303\ RT}{nF} - \log \frac{C_{Ag+}}{C_{AgCl\ (sat)}}$$

$$0.4550 = 0.0591 \log \frac{0.1 \times 0.82}{C_{AgCl}}$$

$$\therefore \quad \frac{0.082}{C_{AgCl(sat)}} = \frac{0.4550}{0.0591} = 7.698$$

$$\therefore \quad \frac{0.082}{\text{Conc of } Ag^+ \text{ in AgCl}} = \text{Ani log } 7.698 = 4.989 \times 10^7$$

$\therefore$ Concentration of Ag^+ in saturated AgCl $= \dfrac{0.082}{4.989 \times 10^7}$

$= 1.644 \times 10^{-9}$ gm ion/litre.

The concentration of Cl^- in KCl

$= 0.1 \times 0.76 = 0.076$ gm ion, litre.

$\therefore$ Solubility product of AgCl at 25°C

$= [Ag^+]\ [Cl-] = 1.644 \times 10^{-9} \times 0.076$

$= 1.249 \times 10^{-10}$

Hence solubility of AgCl at 25°C

$= \sqrt{\ }$ (Solubility product) $= \sqrt{\ }(1.249 \times 10^{-10})$

$= 1.118 \times 10^{-5}$ gm mole/litre

$= 1.118 \times 10^{-5} \times 143.5$ gm/litre

$= 1.603 \times 10^{-3}$ gm/litre.

Example 4:

The E.M.F. of the cell.

Ag | AgCl in 0.1N KCl | NH_4NO_3 Bridge | $AgNO_3$ 0.1N | Ag is + 0.45 volt at 25°C. 0.1 KCl is 85% dissociated and .1N $AgNO_3$ is 82% ionised. Calculate the solubility product of AgCl.

Solution:

$$\text{E.M.F.} = \frac{2.303\ RT}{nF} \log \frac{C_2}{C_1} = 0.0591 \log \frac{C_2}{C_1}$$

(Valency of reversible $Ag^+ = 1$)

or $$0.45 = 0.0591 \log \frac{C_2}{C_1}$$

Now C_2 = concentration of Ag^+ in 0.1N $AgNO_3 = 0.1 \times 0.82 = 0.082$ and C_1 = conc. of Ag^+ in sat AgCl

$\therefore$ $$0.45 = 0.0591 \log \frac{0.082}{C_1}$$

Whence $C_1 = 2.008 \times 10^{-9}$ gm ion/litre.

Since KCl is 85% dissociated, therefore, the Cl^- ion concentration is given by $Cl^- = 0.1 \times 85$.

$$[Cl^-] = 0.085 \text{ gm ion/litre.}$$

Solubility product of AgCl = $[Ag^+]\ [Cl^-] = 2.008 \times 10^{-9} \times 0.085 = 1.7068 \times 10^{-10}$

Example 5:

The E.M.F. of the cell

Ag | AgCNS(s) KCNS (0.1 M) | $AgNO_3$ (0.1 M|) Ag

is 0.586 volt at 18°C. Neglecting the liquid junction potential find the Ag ion concentration in the thiocyanate solution and hence the solubility product of AgCNS.

Solution:

The E.M.F. of the cell is given by

$$E = \frac{2.303\ RT}{1 \times F} \log \frac{C_2}{C_1}$$

$$= 0.058 \log \frac{C_2}{C_1} \qquad \text{(at 18°C)}$$

Substituting $C_2 = 0.1$ gm ion/litre; E = 0.586 volts, we get

$$0586 = 0.058 \log \frac{0.1}{C_1}$$

Whence C_1 = concentration of Ag^+ in AgCNS = 7.942×10^{-12} gm ion/litre

Also concentration of CNS^- in 0.1M KCNS = 0.1 gm ion/litre.

∴ Solubility product of AgCNS

$$= [Ag^+][CNS^-] = 7.942 \times 10^{-12} \times 0.1 = 7.942 \times 10^{-13}.$$

EXERCISES

1. Define 'ordinary' and 'extraordinary' refractive indices for an uniaxial anisotropic material. For the extraordinary beam the ray and wave normal generally do not coincide. Explain.
2. Unpolarized light falls on two polarizing sheets so oriented that no light is transmitted through the combination. If a third polarizing sheet is placed between them, can light be transmitted? Explain.
3. Explain how a quarter wave plate functions to produce circularly and elliptically polarised light.
4. Explain the working of quarter wave plate.
5. What are the optical devices required to produce circularly polarized light from unpolarized light? Explain how they are used to produce circular and elliptical polarised light.
6. What is a half wave plate? Explain its action on polarized light incident on it with its electric vector E making an angle θ with the optic axis of the half wave plate.
7. Discuss the production and detection of circularly polarized light.
8. Explain the phenomenon of double refraction in calcite crystal.
9. What is double refraction?
10. Describe the construction of a Nicol prism and show how it can be used as a polarizer and analyser.
11. Describe in brief the phenomenon of birefringence. Discuss briefly Huygens' theory of double refraction.
12. Plane polarised light is incident on a piece of quartz cut parallel to the axis. Find the least thickness for which the o-ray and the e-ray combine to form plane polarised light.

    Given that $\mu_e = 1.5533$, $\mu_o = 1.5442$

    and $\lambda = 5 \times 10^{-5}$ cm.

2
Diffraction of Material Particles

INTRODUCTION

The intensity of electrons bouncing in different directions is measured by a small collector, whose angular position can be varied. In the collector a retarding potential just short of the accelerating voltage is applied so that only those electrons which bounce with full initial energy are recorded. In the year 1925 Davisson and Germer reported the results of an experiment on the scattering of a beam of electrons impinging on a single crystal of nickel (Fig. 1). In high vacuum, a beam of electrons is accelerated by a potential of ~50 volts. The beam, narrowed by two diaphragms, falls on the single crystal of nickel normal to the {1 1 1} face. Many of the electrons flow into the crystal, while others bounce back.

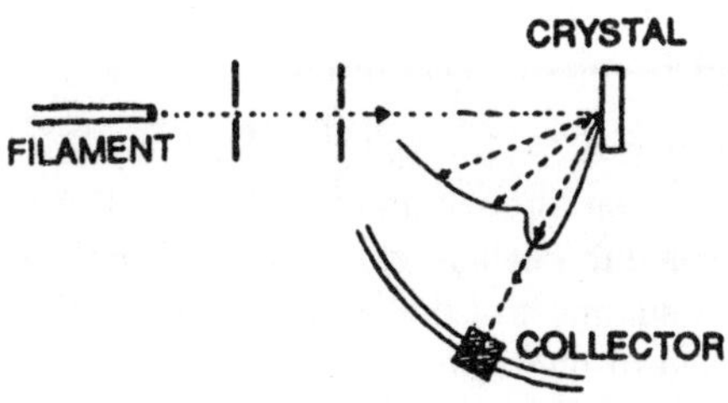

Fig. 1 : Davisson-Germer experiment. The electron currents scattered in various directions are shown by lengths of broken lines. The curved solid line represents the distribution.

OTHER PARTICLES DIFFRACTION

Diffraction experiment's with beams of protons, deuterons and other charged particles have been performed, and all of them establish the validity of the de Broglie relation (1). Due to larger masses in these cases one needs a smaller accelerating voltage to get a wavelength of the proper order. Neutral particles like *neutrons* also show diffraction in the same manner. A velocity-selector sorts out neutrons of a narrow range of momentum P, and special detectors have to be used to observe them.

The unexpected result of the experiment is that the scattered electrons show a pattern with a well-defined maximum and minimum, resembling a diffraction pattern. To explain this, let us tentatively attribute a wavelength λ to the electron beam. As discussed for Bragg reflection, the crystal planes responsible for the maximum must be equally inclined with the directions of incidence and observation.

This gives the grazing angle as $\theta = 90° - 1/2(50°) = 65°$. The nickel crystal had a prominent lattice plane actually at this angle from the beam. Further, the spacing was d = 0.91 A, and using Bragg's equation, the wavelength would be (assuming first order).

$$A = 2 \times 0.91 \times 0.9063 = 1.65.4$$

In 1923 Louis de Broglie had stipulated theoretically that material particles of linear momentum? should show a wave-like behaviour with A given by

$$\lambda = \frac{h}{P} \quad \text{...(1)}$$

where h is Planck's constant. One finds that computation for the 54-volt electron gives $\lambda = 1.67$ A, in close agreement with the value attributed by Davisson and Germer's experiment.

G.P. Thomson in 1927 observed electron diffraction by *transmission* of a fast electron beam through a thin foil of gold. A narrow pencil of *electrons* accelerated through several thousand volts was sent through a gold film so thin that most of the electrons passed out. Now the 'film' is really composed of numerous very small crystals of gold, so that it acts like a 'powder' and with X-rays it produces a ring-shaped diffraction pattern.

WAVE FUNCTIONS

In quantum mechanics, various atomic parameters are described with the help of wave functions ψ. In general, these wave functions ψ are mathematical functions that depend on the variables necessary to specify the particular features of the particle. A wave function is a complex quantity and is not an *observable* quantity; hence it has no direct physical significance. For example, let the location of an electron be described by the wave function ψ (x, y, z, t).

This wave function by itself cannot indicate us the location of the electron. It is known that the intensity of a wave motion is proportional to the square of the amplitude of the wave. Therefore, the square of the absolute value $|\psi|^2$ of the wave function gives where the intensity of the field $|\psi|^2$ is large and thus the regions of space where the particle is more likely to be found at time ψ. The function y is, therefore, called the *probability amplitude*. Thus,

Probability, P, of finding the particle in an infinitesimal volume dV(= dx dy dz) is proportional to $|\psi(x, y, z)|^2$ dx dy dz at time t.

$$\text{or} \quad P \propto |\psi(x, y, z)|^2 \, dV \qquad ...(1)$$

The probability is a real quantity.

The probability of finding the particle between the limits x = a and x = b, y = g and y = h and z = p and z = q, is then

$$P \propto \int_a^b \int_g^g \int_p^q \psi^* (x, y, z)\, \psi (x, y, z)\, dx\, dy\, dz \qquad ...(2)$$

The probability that the electron is located somewhere must be unity.

$$\therefore \quad \iiint \psi^* \psi \, dx\, dy\, dz = 1 \qquad ...(3)$$

SCHRODINGER WAVE EQUATION

In 1926 Erwin Schrodinger developed a wave equation that describes the behaviour of atomic particles. Let us assume that a particle of mass 'm' is in motion along the x-direction. Let the wave function ψ be the dependent variable of the de Broglie wave which is a function of the coordinates x and t. Analogous to the classical wave, we may expect that, ψ will be a function of $(x - \omega t)$. As $\upsilon = \psi/k$, the wave function may be written as a function of $(kx - \psi t)$.

Using the relation $p = \hbar k$ and $E = \hbar\omega$, we can write

$$\psi = f\left(\frac{px - Et}{\hbar}\right) \qquad ...(1)$$

The more general wave would be a sum of a sine and cosine waves. Taking help of Euler's identity, we write the above equation in an exponential form as follows:

$$\psi = A \exp\left[\frac{i}{\hbar}(px - Et)\right] \qquad ...(2)$$

We assume that the energy and momentum of the particle are constant. Differentiating the above equation with respect to x, we get

$$\frac{\partial \psi}{\partial x} = A\,\frac{ip}{\hbar}\exp\left[\frac{i}{\hbar}(px - Et)\right]$$

$$\therefore \quad \frac{\partial \psi}{\partial x} = \frac{ip}{\hbar}\psi$$

Rearranging the terms in the above equation, we get

$$p\psi = \frac{\hbar}{i}\frac{\partial \psi}{\partial x} = -i\hbar\frac{\partial \psi}{dx} \qquad ...(3)$$

Differentiating the eq. (2) with respect to t gives

$$\frac{\partial \psi}{\partial t} = -\frac{iE}{\hbar}A\exp\left[\frac{i}{\hbar}(px - Et)\right] = -\frac{iE}{\hbar}\psi$$

Rearranging the terms in the above equation, we get

$$E\psi\,\frac{\hbar}{i}\frac{\partial \psi}{\partial t} = i\hbar\,\frac{\partial \psi}{\partial t} \qquad ...(4)$$

The partial derivatives with respect to x and t are connected by means of the relation between the energy and momentum. The classical expression for the kinetic energy in terms of the momentum is

$$E_k\,\frac{m\upsilon^2}{2} = \frac{(m\upsilon^2)}{2m} = \frac{p^2}{2m} \qquad ...(5)$$

The total energy and momentum are related by the expression

$$\frac{p^2}{2m} + V = E \qquad ...(6)$$

where Vis the potential energy of the particle. Multiplying the eqn. (5) with w, we obtain

$$\frac{p^2}{2m}\psi + V\psi = E\psi$$

Using the relations (3) and (4) into the above equation, we get

$$-\frac{\hbar^2}{2m}\frac{\partial^2 \psi}{\partial x^2} + V\psi = i\hbar\,\frac{\partial \psi}{\partial t} \qquad ...(7)$$

The above equation is known as the *time-dependent Schrodinger wave equation.*

Knowing the form of V, eqn. (6) can be solved for the wave function ψ. In a number of cases the potential energy V of a particle does not depend on time; it varies with the position of the particle only and the field is said to be *stationary*.

In the stationary problems Schrodinger equation can be simplified by separating out time and position- dependent parts. Accordingly, we can write the wave function as a product of x, $\psi(x)$ and a function of t, φ (t).

We, therefore, write that

$$\psi (x, t) = \psi (x) \varphi (t) \qquad ...(8)$$

Equation (6) may be written as

$$-\frac{\hbar^2}{2m} \varphi \frac{\partial^2 \psi}{\partial x^2} + V \psi \varphi = i\hbar \psi \frac{\partial \psi}{\partial t}$$

Dividing the above equation with ψ_{φ}, we get

$$-\frac{\hbar^2}{2m} \frac{1}{\psi} \frac{d^2 \psi}{\partial x^2} + V = i\hbar \frac{1}{\varphi} \frac{d\varphi}{\partial t} \qquad ...(9)$$

If we assume that the potential energy Vis a function of x only, the entire left hand side of eqn. (8) is a function of x only while the right hand side is a function of t only. Since x and t are independent variables, both the function of x and t must be equal to a constant. The constant that each side must equal is called the *separation constant* E. Thus,

$$-\frac{\hbar^2}{2m} \frac{1}{\psi} \frac{d^2 \psi}{\partial x^2} + V = E \qquad ...(10)$$

and $$i\hbar \frac{1}{\varphi} \frac{d\varphi}{\partial t} = E$$

Eq. (9) may be rewritten as

$$-\frac{\hbar^2}{2m} \frac{2^2 \psi}{dx^2} + V\psi = E\psi$$

or $$\frac{2^2 \psi}{dx^2} + \frac{2m}{\hbar^2} (V - E) \psi = 0$$

$\therefore$ $$\frac{2^2 \psi}{dx^2} + \frac{8\pi^2 m}{h^2} (V - E) \psi = 0 \qquad ...(11)$$

The above equation is called the time-independent Schrodinger equation.

ALLOWED WAVE FUNCTIONS AND ENERGIES

The time-independent wave equation is the pertinent equation for studying properties of atomic systems in stationary conditions. Wave mechanical methods of solving the problem of particle motion are essentially based on ψ functions. Appropriate wave equation is formulated by incorporating the particle mass 'w' and potential energy function V for the region in which the particle is located. The next step consists of solving the differential equation for solutions, namely for the ψ functions which will satisfy the differential equation.

By solving the Schrodinger equation, we obtain the possible set of ψ functions. In case of bound particles the acceptable solutions for the differential equation are possible only for certain specified values of energy. These energy values will be the only possible results of precise measurements of the total energy of the particle. These discrete values of energy E_1, E_2, ... E_n are called eigen values or allowed values of the energy of the particle. The solutions ψ_1, ψ_2, ... ψ_n corresponding to the eigen energy values E_n are called the eigen functions. The quantization of energy thus appears as a natural element of the wave equation.

Thus, using the Schrodinger equation, it is possible to determine first the electron energies and then the wave functions. These wave functions can then be used to determine the probability distribution function $\psi^*\psi$ of the electron, for various discrete energies as it revolves around the nucleus.

OPTICAL INDICATRIX

The different speeds of light in different directions in a crystal are conveniently represented by a three-dimensional figure that shows the refractive index for light waves in their direction of polarization. The representation is known as the *indicatrix*. *The optical indicatrix is spherical or ellipsoidal surface having the three different refractive indices as axes*. In case of isotropic crystals, the axes are all of the same length and consequently the refractive indices in three directions are the same. Therefore, the indicatrix is a *sphere*. In case of uniaxial crystals, a unique axis exists normal to the plane of two equal axes.

The velocity of propagation of light along this unique axis (z-direction) is independent of the wave polarisation and this direction is the *optic axis* direction. The refractive index along this direction is μ_e. In a plane perpendicular to the optic axis, the refractive index is independent of direction, that is $\mu_x = \mu_y$ denoted by μ_0. Therefore, in uniaxial crystals the indicatrix is an ellipsoid having a circular cross-section in the xy-plane. The ellipsoid is prolate if $\mu_e > \mu_0$ and oblate if $\mu_e < \mu_0$.

The form of optical indicatrix can be obtained as follows. The energy density in a dielectric medium is given by

$$W = \frac{1}{2} D.E \qquad ...(1)$$

$$W = \frac{1}{2}\left[\frac{D_x^2}{\varepsilon_{11}} + \frac{D_y^2}{\varepsilon_{22}} + \frac{D_z^2}{\varepsilon_{33}}\right] \qquad ...(2)$$

or $$\frac{1}{2}\left[\frac{D_x^2}{W\varepsilon_{11}} + \frac{D_y^2}{W\varepsilon_{22}} + \frac{D_z^2}{W\varepsilon_{33}}\right] = 1$$

or $$\left[\frac{D_x^2}{2W\mu_x^2\varepsilon_0} + \frac{D_y^2}{2W\mu_y^2\varepsilon_0} + \frac{D_z^2}{2W\mu_z^2\varepsilon_0}\right] = 1$$

Writing $x^2 = \frac{D_x^2}{2\varepsilon_0 W}$, $y^2 = \frac{D_y^2}{2\varepsilon_0 W}$ and $z^2 = \frac{D_z^2}{2\varepsilon_0 W}$, we get

$$\frac{x^2}{\mu_x^2} + \frac{y^2}{\mu_y^2} + \frac{z^2}{\mu_z^2} = 1 \qquad ...(3)$$

This is the equation of an ellipsoid with semi-axes μ_x, μ_y and μ_z. In the case of a uniaxial crystal the indicatrix has circular symmetry about the z-axis. In this case,

$$\frac{x^2}{\mu_0^2} + \frac{y^2}{\mu_0^2} + \frac{z^2}{\mu_e^2} = 1 \qquad ...(4)$$

Let us now consider light propagating in a direction r, at an angle θ to the optic axis. Because of the circular symmetry, we can choose that the y-axis should coincide with the projection of r on the xy-plane. The plane normal to r intersects the ellipsoid in the shaded ellipse. The two allowed directions of polarization are parallel to the axes of the ellipse and thus correspond to OP and OQ (Fig. 2). They are thus perpendicular to r as well as to each other. The two waves polarized along these directions have refractive indices given by OP = μ_o and OQ = $\mu_e(\theta)$. In case of the e-ray, the plane of polarization varies with θ as does the refractive index.

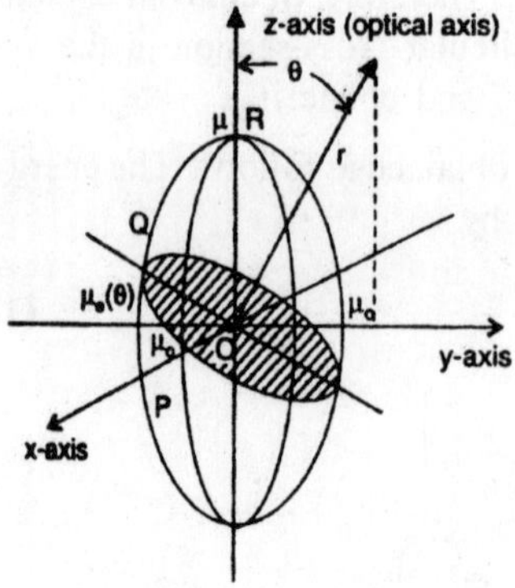

Fig. 2

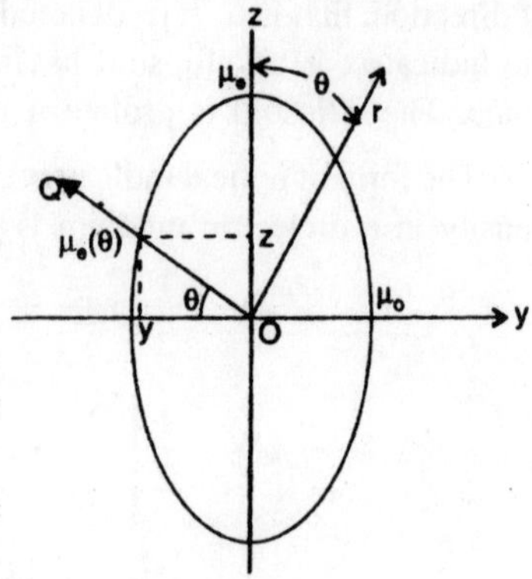

Fig. 3

We can determine the relationship between $\mu_e(\theta)$, μ_e and μ_o with the help of the Fig. 3. The figure shows the intersection of the indicatrix with the yz-plane. From the diagram, it is seen that

$$\mu_e^2(\theta) = z^2 + y^2 \quad \text{...(5)}$$

and $$z = \mu_e(\theta) \sin\theta \quad \text{...(6)}$$

$\therefore$ $$\mu_e^2(\theta) = \mu_e^2(\theta)\sin^2\theta + y^2$$

or $$y^2 \; \mu_e^2(\theta)\cos^2\theta \quad \text{...(7)}$$

The equation of the ellipse shown in Fig. 3 is

$$\frac{y^2}{\mu_o^2} + \frac{z^2}{\mu_e^2} = 1 \quad \text{...(8)}$$

Substituting eqn. (6) and eqn. (7) into eqn. (8), we obtain

$$\frac{1}{\mu_e^2(\theta)} = \frac{\cos^2\theta}{\mu_o^2} + \frac{\sin^2\theta}{\mu_e^2} \quad \text{...(9)}$$

Thus, for $\theta = 0°$, that is propagation along the optic axis, $\mu_e(0°) = \mu_o$ while for $\theta = 90°$, $\mu_e(90°) = \mu_e$. The two polarizations which can be propagated correspond to the maximum and minimum refractive indices given by the index ellipsoid. For propagation parallel to the optic axis (that is z-direction), there is no birefringence as the section of the ellipsoid perpendicular to this direction is a circle. For propagation perpendicular to the optic axis, the birefringence will be a maximum, the permitted polarizations will be parallel to the y-axis with refractive index μ_0 and parallel to the z-axis with refractive index μ_e.

ELECTROMAGNETIC THEORY OF DOUBLE REFRACTION

The birefringent crystals belong to the group of non-conducting materials. They are known, as dielectric materials. If a dielectric crystal is placed between the plates of a parallel plate capacitor and a voltage is applied, the crystal becomes polarized. This polarization is dielectric polarization and is different from the optical polarization.

The voltage produces an electric field E and under the action of the electric field, electron clouds in the atoms of the crystal distort with the result that the centres of action of negative charge and positive charge get separated by a small distance. It means that electrical dipoles are produced throughout the crystal. Negative charges of the dipoles lie on the side of the positively charged plate of the capacitor and positive charges lie on the side of the negatively charged plate.

Thus electric charges of opposite nature are induced on the opposite faces of the crystal in response to the applied electric field (Fig. 4). This is known as *dielectric polarization*.

The electric field, polarization and displacement in the crystal are related by

$$D = \varepsilon_0 + P \qquad ...(1)$$

where $$D = \varepsilon_0 \varepsilon_r E \qquad ...(2)$$

and $$P = \chi \varepsilon_0 E \qquad ...(3)$$

ε_r and χ are known as *relative permittivity* and *electric susceptibility* of the crystal. The relative permittivity is related to the refractive index of the material through the relation

$$\mu = \sqrt{\varepsilon_r} = \sqrt{\frac{\varepsilon}{\varepsilon_0}} \qquad ...(4)$$

We know that refractive index is related to the velocity of light in the medium.

$$\mu = \frac{c}{\upsilon} \qquad ...(5)$$

It is not necessary for dielectric polarization that the crystal be placed in between the plates of a capacitor. The electric field of a light wave propagating in the crystal also can cause dielectric polarization. According to the above relations the dielectric polarization caused by the light wave is related to the velocity (direction of propagation) of light wave with in the crystal.

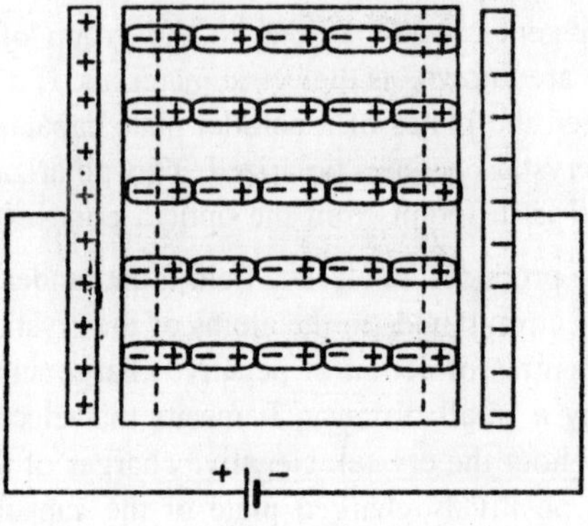

Fig. 4 : Polarization of a dielectric

In case of isotropic crystals, the vectors D, E and P are parallel and ε_r, μ and χ are scalars. In anisotropic crystals the induced polarization is in a direction different from that of the field. As a result, ε_r, μ and χ are tensor quantities. The components of polarization are given by

$$P_x \; \varepsilon_0 \, (\chi_{11} \, E_x + \chi_{12} \, E_y + \chi_{13} \, E_z)$$
$$P_y \; \varepsilon_0 \, (\chi_{21} \, E_x + \chi_{22} \, E_y + \chi_{23} \, E_z) \quad \text{...(6)}$$
$$P_z \; \varepsilon_0 \, (\chi_{31} \, E_x + \chi_{32} \, E_y + \chi_{33} \, E_z)$$

The above equations can be simplified by choosing the co-ordinate axes in such a way that the off-diagonal elements vanish. Then

$$P_x \; \varepsilon_0 \, \chi_{11} \, E_x$$
$$P_y \; \varepsilon_0 \, \chi_{22} \, E_y \quad \text{...(7)}$$
$$P_z \; \varepsilon_0 \, \chi_{33} \, E_z$$

Similarly, we can express $D_x = \varepsilon_{11} \, E_x$,

$$D_y = \varepsilon_{22} \, E_y \quad \text{...(8)}$$
$$D_z = \varepsilon_{33} \, E_z$$

These directions are called the *principal axes* of the crystal and the corresponding diagonal terms ε_{11}, ε_{22} and ε_{33} the *principal permittivities*. The variation in the permittivity and the corresponding variation in refractive index and hence in wave velocity is the origin of double refraction. μ_x, μ_y and μ_z are the three principal *refractive indices* corresponding to the three principal permittivities. *The subscripts x, y and z relate to the direction of polarization of the light waves and not to their direction of propagation.* The principal permittivities and refractive indices are related as

$$\mu_x = \sqrt{\frac{\varepsilon_{11}}{\varepsilon_0}}, \mu_y = \sqrt{\frac{\varepsilon_{22}}{\varepsilon_0}} \text{ and } \mu_z = \sqrt{\frac{\varepsilon_{33}}{\varepsilon_0}} \quad ...(9)$$

DIFFRACTION OF NEUTRON AND ELECTRON

The theoretical significance of wave-nature exhibited by material particles will be treated in the next chapter. About the use of electron beams and ion beams in microscopy we have spoken 15.8. They are now widely used to see the details of the order of a few angstroms—close to molecular sizes.

Here we will discuss the use of electron diffraction and neutron diffraction for studying the crystalline structure of materials. The procedure is the same as with X-rays, but fortunately these methods have complementary character. Let us see this.

X-ray interaction occurs through electromagnetic nature of light atoms on the Z electrons of the atom. This interaction is weak; therefore the X-rays penetrate deep, and X-ray diffraction gives information about the entire interior of a solid. The information is more about the higher Z atoms (like Pb, V, etc.) than about the lighter atoms (like H, C, O, etc.).

In contrast, interaction of electrons occurs through electrostatic force This being a much stronger interaction the beam does not peneyate deep This is not a disadvantage always, because if you want to study the structure of thin films, or surface films on a solid, or even the structure of liquid near the surface (which can be different from that in the interior), then electron diffraction is the best. Electrons also have a magnetic moment, and therefore its behaviour also depends on the magnetic character of the surface layers. That is a great advantage.

Neutrons have no charge and their penetration in the solid is very deep-even more than X-rays. But the interaction of neutrons does not depend on Z; it depends on the fact that it has a magnetic moment μ. So if there are isotopes with different μ, or neighbouring Z atoms with different μ then neutron scattering experiments will be able to distinguish between them. Another fact is that, for a given interaction strength, the momentum sharing is more in collisions with particles of comparable mass (like H, D, Be, He etc.), than with heavier or much lighter ones.

For this reason neutron diffraction gives far more information about the nuclei of the lighter atoms in the crystal lattice than what X-ray. diffraction does.

Thus, for material in bulk X-ray diffraction gives best information for high Z atoms and ignores the effect of μ of the target atoms/ions. Neutron diffraction gives information dominantly depending on μ of the nuclei of target ions/atoms, and more so for the lightest atoms. Elęctron diffraction gives no information about the bulk material, but about thin films and surface layers, it is the only method. The complimentary features make a combination so useful.

PROPERTIES OF SPECTRAL LINES

We record me spectral lines with the help of spectrometers, which have narrow slits. The spectral lines are therefore linear and possess three important properties; namely *position* measured in terms of frequency, finite *width*, which is a *range* of frequencies and *intensity*, the brightness of the line.

Width of Spectral Lines

Real atoms or molecules do not emit, radiation at precise frequencies; each emission is more or less broadened by various processes, and so each line is really a small package of slightly different frequencies. Therefore, even if we make the slit of the spectrometer infinitely narrow, there is nonetheless a minimum line width. The natural shape of an emission line appears under ideal conditions when the emitting atom is at rest and is not subjected to the external forces during the process of emission. In practice, several factors contribute to the line broadening. They are broadly classified into two categories:

(i) Homogeneous and

(ii) Inhomogenenous broadening. If the broadening mechanism affects each individual atom in the sample to the same extent, then the broadening is said to be *homogeneous*. In such a case all of the atoms in the sample will have the centre frequency v_0 and the same line shape. Natural broadening and collision broadening belong to this category. On the other hand, if different atoms in the sample have slightly different frequencies for the same transition, the overall response of the sample broadens out and the broadening is said to be *inhomogeneous*. Doppler broadening and broadening due to crystal defects belong to this category.

1. *Collision broadening* : Atoms or molecules in liquid and gaseous phases are in continual motion and collide frequently with each other. When a radiation-emitting atom undergoes a collision, the emission conditions change and it is equivalent to an interruption. In the process, the phase of the emitted wave suffers random variations.

2. *Doppler broadening* : Doppler effect occurs when a source and an observer are in relative motion. The frequency as measured by the observer increases if the source and observer approach each other and decreases when they recede. In a gas due to the chaotic nature of thermal motion, all the directions of the molecules' velocities relative to a spectrograph are equally probable. Therefore, the radiation received by the spectrograph contains all the frequencies in the interval from $v_0 (1 - \upsilon/c)$ to $v_0 (1 + \upsilon/c)$, where v_0 is the frequency emitted by the molecules and υ is the velocity of thermal motion. Hence the spectral line is broadened. The broadening of the line caused by the Doppler effect is called *Doppler broadening.* In general, for liquids collision broadening is the most important factor, whereas for gases where collision broadening is less pronounced, the Doppler effect often determines the natural line width.

3. *Natural broadening* : Even in an isolated, stationary atom the energy levels are not indefinitely sharp. According to Heisenberg uncertainty principle, if a system exists in an energy state for a limited time Δt seconds, then the energy of that state will be uncertain to an extent ΔE where

$$\Delta E \times \Delta t \approx h/2\pi$$

Thus, we see that the ground state of a system is sharply defined since, left to itself, the system will remain in that state for an infinite time. Thus $\Delta t = \infty$ and $\Delta E = 0$. In contract, the lifetime of an excited state is about 10^{-8}s, which gives a value for ΔE of about 10^{-34}J.s $\div$ 10^{-8} s $= 10^{-26}$ J. A transition between an excited state and the ground state will thus have an energy uncertainty of ΔE, and a corresponding uncertainty in the associated radiation frequency of $\Delta E/h$, which we can write as

$$\Delta v = \frac{\Delta E}{h} \approx \frac{1}{2\pi \Delta t} \qquad ...(1)$$

If an excited electronic state lifetime is 10^{-8}s, then $\Delta v = 10^8$Hz. Thus, the natural broadening is relatively small in magnitude and it is often masked by other mechanisms.

THE INTENSITY OF SPECTRAL LINES

The spectral line intensities are dependent on two factors: the *transition probability*, the likelihood of a system in one state changing to another state; and the *population of state*, the number of atoms or molecules initially in the state from which the transition occurs.

1. *Transition probability* : The detailed calculation of absolute transition probabilities involves a knowledge of the precise quantum mechanical wave functions of the two states between which the transition occurs. It is often possible to decide whether a particular transition is allowed or forbidden basing on selection rules.
2. *Population of states* : If we have two levels from which transitions to a third are equally probable, then obviously the most intense spectral line will arise from the level which initially has the greater population. At thermal equilibrium the population of a set of energy levels is governed by the Boltzmann law

$$\frac{N_2}{N_1} = e^{-\Delta E/kT} \quad ...(1)$$

where N_1 is the population in the lower energy state E_1 and N_2 in the higher energy state E_2, $\Delta E = E_2 - E_1$, T is the temperature in K, and k is Boltzmann's constant.

LUMINESCENCE

A body that emits light on account of high temperature is called *incandescent*. There are bodies that emit light due to causes other than high temperature and they are said to be *luminescent*. The emission of light from a substance, when it is stimulated by the input of energy from a source of suitable radiation, is known as luminescence. Materials exhibiting *luminescence* are broadly called as *phosphors*. There are various ways of exciting substances and so there are the following kinds of luminescence.

(i) *Photoluminescence* is the glow of bodies as a result of their irradiation by visible or uv light, x-rays or γ-rays. Tube lights use this phenomenon to produce visible light. The internal surface of the tube is coated with phosphors and these substances emit visible light of lower frequency under the action of uv light of higher frequency.

(ii) *Electroluminescence* is caused due to the passage of an electric current through a substance or the action of an electric field. The glow of a gas discharge in the tubes of advertisement is an example of this kind of luminescence.

(iii) *Cathodoluminescence* is the glow of materials due to the bombardment of the material by electrons or other charged particles.

(iv) *Chemiluminescence* is caused at the expense of energy produced in a chemical reaction. The glow of many living organisms such as bacteria and insects is due to chemiluminescene.

Many substances emit light when a beam of light is made to fall on them. When the light is emitted only as long as the exciting radiation is maintained, it is called *fluorescence*. For fluorescence, the time interval between the acts of excitation and emission is 1 to 10 ns. For example, kerosene emits a faint bluish fluorescence when illuminated by daylight.

Substances, which continue to emit light for some time after the exciting radiation is removed, are called *phosphorescent* substances. Phosphorescence is displayed for example by zinc sulphide. A coat of zinc sulphide applied to a sheet of glass makes a fluorescent screen, which will emit a faint glow for minutes after the exciting radiation is switched off.

Fluorescence occurs when the atoms of the fluorescent material absorb a portion of the incident radiation of shorter wavelength and re-emit it in the form of radiation of longer wavelength.

Luminescence in solids is closely related to impurities and lattice defects. Crystal luminescence is more important and practically useful. Whatever may the form of energy given to the luminescent material, the final stage in the process is an electronic transition between two energy levels, E_1 and E_2 with the emission of light of frequency v, where

$$v = \frac{E_2 - E_1}{h}$$

Invariably E_1 and E_2 are part of two groups of energy levels, so that, instead of a single emission frequency, a band of frequencies is usually observed.

When an electron is excited from the valence band into the conduction band, a hole is left in the valence band. In a perfectly regular lattice, the electron usually returns to the valence band, although it may take some time to do so. However, if the lattice has some impurity that introduces energy levels in the forbidden region, an electron in a low lying impurity level may fill the hole in the valence band while the electron in the conduction band may fall into one of the vacant high impurity energy levels. These transitions generally involve photons of small energy that do not fall into the visible region. Finally, the electron may fall from higher energy impurity level to the lower energy impurity level, emitting radiation of lower energy or longer wavelength than the incident radiation; this constitutes the luminescence.

In some instances, instead of going through the process depicted, the electron in the conduction band may fall into an energy level called a *trap*. From there electron cannot jump to the ground state impurity energy level. The transition is forbidden. In such a case the electron is in a state similar to an atom in a metastable state. The trapped electron waits till it is returned to the conduction band through some other mechanism. Then it follows the steps (b) and (c). Due to the time delay involved, the process is *phosphorescence*. These substances are called *phosphors*. One such substance is zinc sulphide. Phosphors are widely used in coating the screen of CRTs used in CROs, TVs and video terminals of computers.

QUANTUM THEORY

The explanation of Raman scattering in terms of the quantum theory is very simple. When light is incident on a solid, liquid, or gas, the photons can be imagined to undergo collisions with molecules.

(i) If the collision is perfectly *elastic*, there will be no transfer of energy from the photon to the molecule or from the molecule to the photon. The photon is scattered without any change of energy. Therefore, the frequency of the scattered photon is the same as that of the incident photon. A detector placed to collect energy at right angles to an incident beam will thus receive photons of energy hv, that is the radiation of frequency V. This explains the Rayleigh line in the Raman spectrum.

(ii) However, it may happen that energy is exchanged between photon and molecule during the collision; such collisions are called *inelastic collisions*. The molecule can gain or lose amounts of energy only with the accordance with the quantum laws. That is its energy changes must be the difference in energy between two of its allowed states. That is to say, ΔE must represent a change in the vibrational and/or rotational energy of the molecule.

If the molecule gains energy ΔE (= $E_2 - E_1$) from the photon, it goes from a lower energy level E_1 to a higher energy level E_2, while the photon will be scattered with energy ($h\nu - \Delta E$). Then, the scattered radiation will have a frequency ($\nu - \Delta E/h$) which is less than that of the incident radiation. The resulting lines are Stokes lines located on the lower frequency side of the Raman spectrum.

Conversely, the molecule may be initially in an excited state. It may lose energy ΔE to the photon and go to a lower state after collision, while the photon is scattered with energy ($h\nu + \Delta E$). Hence, the scattered radiation will have a frequency ($\nu + \Delta E/h$). The resulting lines are anti-Stokes lines, which are situated on the higher frequency side of the Raman spectrum.

At ordinary temperatures, there are more molecules in the lower energy state. Therefore, transitions are more likely from lower energy state to upper energy state. Hence, energy is absorbed by the molecules from the photons. The reverse process of energy being given to photons by molecules is less likely. Therefore, Stokes lines are more intense than the anti-Stokes lines. The intensity of anti-Stokes lines increases with temperature, as a rise in temperature increases the relative population of molecules in the higher energy state.

FOURTEEN BRAVAIS SPACE LATTICES

All three have a cubic shape. The first has lattice points at the corners only; the second has an extra lattice point at the centre of *each* face; the third has an extra point at the centre of its volume. Consider the three unit cells shown in Fig. 5. These fall under *one system* (cubic system), but *three different space lattices*.

Bravais showed that there could be in all 14 different lattices under the seven systems. When there is no extra lattice point in the unit cell besides the corners, the cell is called a *primitive cell* (P). When there are lattice points at the face-centres or at the body centre, symbols F and I respectively are used. Symmetry considerations decide which of these can be had in each system. For instance, in the cubic system" we have P, F and I lattices. In triclinic system only primitive cells (P) can occur.

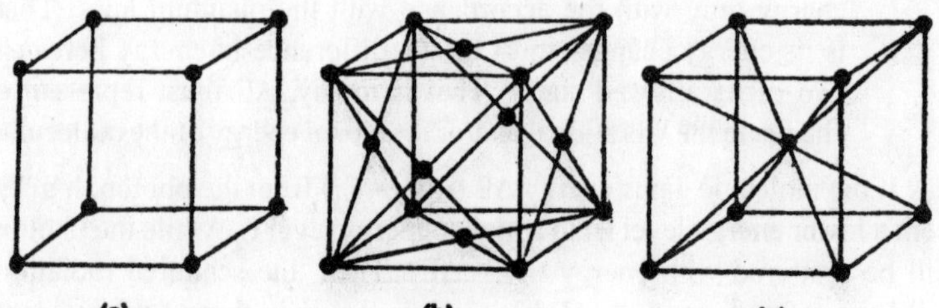

Fig. 5 : Unit cells in simple cubic (cubic P), face-centred cubic (cubic F) and body-centred cubic (cubic I) space lattices of Bravais.

Cubic P, cubic F, and cubic I lattices are also referred to as sc (simple cubic), *fcc* (face-centred cubic) and *bcc* (body-centred cubic) lattices respectively. One way of looking at the bcc lattice is to consider it as one *sc* lattice plus one more shifted by $\left(\frac{1}{2}, \frac{1}{2}, \frac{1}{2}\right)$. Similarly *fcc* lattice is equal to one *sc* lattice plus *three* more shifted by $\left(0, \frac{1}{2}, \frac{1}{2}\right)\left(\frac{1}{2}, 0, \frac{1}{2}\right)$ and $\left(\frac{1}{2}, \frac{1}{2}, 0\right)$. [All coordinates here are expressed in units of a].

DOUBLY REFRACTING CRYSTALS

Certain crystals, such as quartz and calcite, as well as many others (including ice, to a very slight extent) have different properties in different directions.

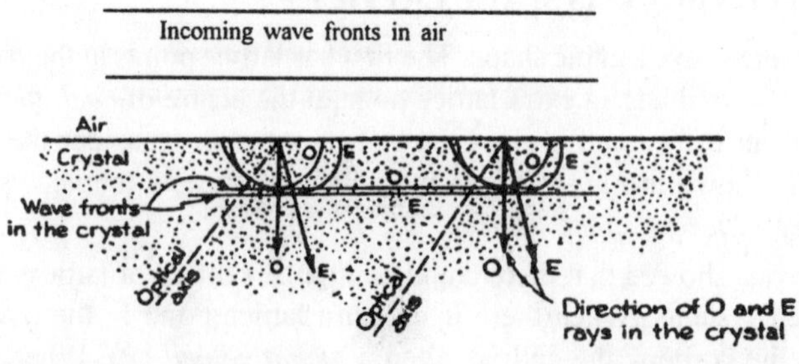

Fig. 6 : Spreading of light waves in a doubly refracting crystal. The "ordinary" waves spread out in spherical waves, the "extraordinary" waves in elliptical waves. In the direction called the "optical axis," the speed of both ordinary and extraordinary waves are the same.

Electrical properties, heat conductivity, and optical properties depend on the direction in which they are measured in the crystal. Fig. 6 shows the spreading of light waves in such a crystal.

In one particular direction, called the *optical axis* ("axis" here means a direction, and not a particular line), all light travels at the same speed, no matter what its plane of vibration. Ray A in Fig. 7 indicates a ray parallel to the optical axis of a crystal, in which all components of vibration have the same velocity. Ray B is perpendicular to the optical axis direction, and all of the vibrations in B can be resolved into components that are either parallel to the optical axis (E) or perpendicular to it (O). The O components (called the "ordinary" ray) travel at the same speed as A; the E components (called the "extraordinary" ray) travel at a different speed. (In Fig. 7 they are shown with a higher speed than O, but in some crystals the speed may be slower).

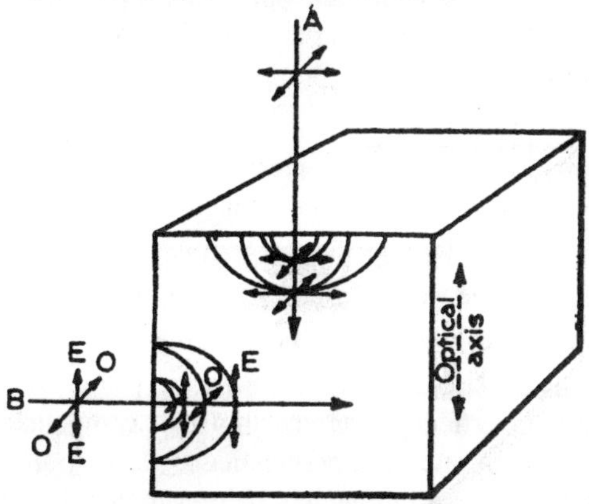

Fig. 7 : In ray A, parallel to the optical axis, vibrations ore all per-pendicular to the optical axis, and all travel at the same speed. Ray B is perpendicular to the optical axis, and its vibrations will hove components both perpendicular and parallel to the optical axis. The components perpendicular to the optical axis travel at the normal speed, and comprise the "ordinary" ray. The components vibrating parallel 10 the optical axis (the "extraordinary" ray) travel at a different speed.

We have defined the index of refraction, v, to be the ratio of the speeds of light in air (or vacuum) and in the material. A doubly refracting crystal will have two indices of refraction: re for the extraordinary ray and v_0 for the ordinary ray. Indices of refraction for the ordinary and the extraordinary ray are given here for a few commonly used materials:

	v_0	v_E
Calcite	1.658	1.486
Quartz	1.544	1.553
Tourmaline	1.637	1.619

Tourmaline has the property of being relatively opaque to the ordinary ray, so that only the extraordinary is transmitted through it. Thus, if a ray of unpolarized light-falls on a tourmaline crystal in a direction perpendicular to its optical axis, the light coming through the crystal will emerge completely polarized—*i.e.*, with all the vibrations in a single plane that is parallel to the optical axis. Iodo-quinine sulfate crystals have this same property of absorbing one ray and being transparent to the other. "Polaroid" is composed of billions of tiny needle-shaped iodoquinine sulfate crystals lined up with their optical axes parallel and imbedded in a sheet of plastic. Light passing through a sheet of Polaroid is almost completely polarized, with the vibrations being parallel to the length of the imbedded crystals.

The production of colors by means of doubly refracting material gives an interesting insight into the behaviour of polarized light.

(1) Falling on a piece of polaroid with a vértical axis. The light emerging from this polarizer is all vibrating in a vertical plane,

(2). The beam next falls on a slab of doubly refracting crystal with its optical axis 45° from the vertical, The crystal resolves die incoming vertical vibrations into two components, E and O, which are parallel and perpendicular to the optical axis of the crystal. These vibrations will be transmitted through the crystal at different speed and, if the thickness of the crystal is Just right, the extraordinary ray will gain a half wave length on the ordinary ıay by the time they emerge into the air again

(3). Since the vibrations of E and O are at right angles to one another, there can be no interference between them. However, if we now have the beam pass through another piece of polaroid (the analyzer) with an axis 45° to the plane of both E and O, only the components parallel to the analyzer axis can come through.

(4) These components now lie in the same plane of vibration and are able to interfere with one another. Since E and O are a half wave length out or phase and of equal amplitude, they will annul one another and the light of the beam will be extinguished.

We have considered the beam to be monochromatic, and it is plain that if we let white light fall on our apparatus the thickness of the plate may be suitable to extinguish light of some wave lengths (if the retardation is any odd number of half wave lengths) and reinforce others (if the retardation is any integral number of wave lengths). Ordinary cellophane is doubly refracting, and if a wad of crumpled cellophane is placed between two sheets of polaroid, so that the light passes through different thicknesses of cellophane, beautiful colored patterns can be seen.

The production of colors by the above means is used in the study of mineral crystals. Plate III shows a photograph of crystals taken in this way.

ANISOTROPIC CRYSTALS

When a light beam is incident on an *isotropic medium* such as a glass slab, it refracts as a single ray. An optically isotropic material is one in which the index of refraction is the same in all directions. Glass, water and air are examples of isotropic materials. The atoms in a crystal are arranged in a regular periodic manner. If the arrangement of atoms differs in different directions within a crystal, then the physical properties vary with the direction.

The thermal conductivity, electrical conductivity, velocity of light and hence refractive index etc., properties depend on the crystallographic direction along which the property is measured. Then we say that the crystal is *anisotropic*. In such anisotropic crystals the force of interaction between the electron cloud and the lattice is different in different crystallographic directions. The natural frequency of the electron cloud is likewise dependent on the direction in which the electrons are caused to vibrate by the incident light wave. This results in different velocities in different directions and the index of refraction is different in different directions within the crystal.

The anisotropic crystals are divided into two classes: *uniaxial* and *biaxial* crystals. In case of uniaxial crystals, one of the refracted rays is an ordinary ray and the other is an extraordinary ray. In biaxial crystals both the refracted rays are extraordinary rays. Calcite, tourmaline and quartz are examples of uniaxial crystals whereas mica, topaz and aragonite are examples of biaxial crystals.

We study here only about the uniaxial crystals.

X-RAYS DIFFRACTION

When X-rays fall on a crystal, each electron of each atom scatters a part of the beam, and we have to consider the sum-total of the scattered radiation. This may be seen in three steps:

(i) The amplitude of the scattered wave resulting from Z different electrons of a *given atom* is not just Z-fold the amplitude of scattered wave per electron; it is AZ-fold, whereas is called the atomic structure factor. It depends on the distribution of electrons in the atom, A of the X-rays used and the direction of observation relative to the incident light.

(ii) The amplitude of the scattered wave resulting from the different atoms in a unit cell of the crystal is not just $\Sigma A_i Z_i$ summed over the atoms in the unit cell, but is $f\Sigma A_i Z_i$, where f is called the *crystal structure factor*. It depends on the relative placements of the-atoms in a unit cell, on λ of the X-rays used, and on the direction of observation.

(iii) The amplitude of the scattered wave from a crystal as a whole is then the resultant arising from the *lattice arrangement* of different unit cells. For some chosen directions the contributions from all unit cells addup in phase. These form the maxima, and their positions depend on the lattice parameters, the λ value, and direction of observation.

Laue followed the procedure of treating the unit cell as the scatterer and summing up over the crystal as a whole. In chosen directions alone are the scattered beams from the assembly of unit cell in *phase agreement*, and in these directions, the maxima occur as *Laue spots*. Their *positions* and *intensities* lead us to (he lattice arrangement and the structure details within each unit cell. In fact, even the charge density distribution within atoms is obtained.

But we will use a simpler (limited) procedure used by Bragg, who centered attention on *atomic planes*.

Bragg's Law : In Fig. 8 we show a set of parallel atomic planes of a crystal with spacing a. The wave-normal of the incident beam is OP at *glancing angle* θ (that is the practice in X-rays), and the direction of observation is A_1Q_1 at glancing angle θ' on the other side. The atoms (in the planes) are not shown, because the result is in dependent of their locations within a given plane.

Now, if $\theta' = 0$ corresponding to specular reflection (*i.e.* A_1Q_1 also lies in the plane of incidence) then it is simple to see that the scattered waves from all the atoms in a given plane reach the observer in the same phase, and therefore reinforce.

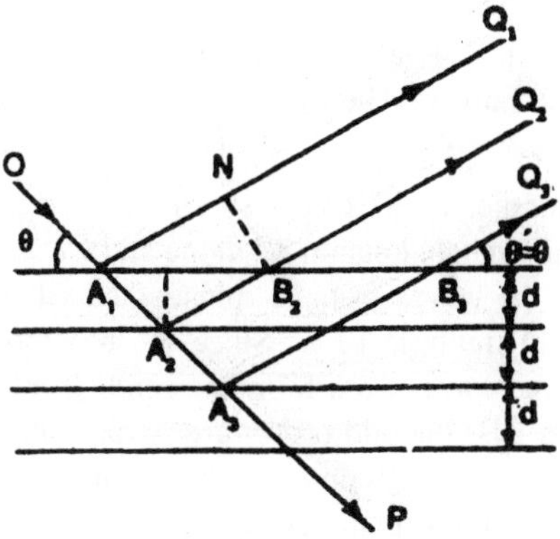

Fig. 8 : Brass's Law.

The total scattered waves from successive planes have a path difference

$$p = A_1A_2B_2 - A_1N$$

where N is the foot of perpendicular from B_2 on A_1Q_1. We have

$$p = \frac{2d}{\sin\theta} - 2d\cot\theta\cos\theta = \frac{2d}{\sin\theta}(1-\cos^2\theta) = 2d\sin\theta$$

For reinforcement this must be an integral multiple of λ. Hence

$$2d\sin\theta = n\lambda \qquad ...(1)$$

Thus specular reflection condition *plus* Bragg's law means that waves *from all unit cells in all the lattice planes* reach the observer in agreement of phase.

An ordinary mirror behaves with visible light exactly the same way—each atom of the mirror being the centre of scattering. But for visible light the penetration is not large and $\lambda >> d$. Hence Bragg's condition plays no role and light of all wavelengths shows a maximum at $\lambda' = \lambda$.

If the crystal belongs to a compound, then the planes, may be treated as belonging to one kind of atoms. Eqn. (1) gives the maxima so far as contribution from one kind of atoms are concerned. The atomic planes for other kinds of atoms have a relative shift. We shall consider only a diatomic crystal. Just two kinds of situation arise:

(i) for some (h k *l*) values the planes may cover both the kinds of atoms, and

(ii) for other (h k *l*) values the planes containing the other kind of atoms lie half-way between those containing the first kind of atoms.

As one example, in KC*l* (whose structure is of NaCl type) the {1 1 1} planes alternately contain K^+ alone and Cr^- alone. Therefore for odd values of n the waves from K^+ planes and Cl^- planes have phase difference an odd multiple of π. Since K^+ has the same number of electrons as CF the waves from K^+ planes and Cl^- planes totally cancel out for orders. In KBr the odd orders are weaker than even orders. But {1 0 0} plane for example, contain both K^+ and Cl^- (or Br^-) ions, and for these planes all the orders appear.

VELOCITY OF LIGHT

The first attempt to measure the propagation of light was undertaken by Galileo in a very primitive way. One evening, he and his assistant placed themselves on two distant hills in the neighbourhood of Florence, each of them carrying a lantern with a shutter. Galileo's assistant was instructed to open his lantern as soon as he noticed the flash from the one carried by his master. If light was propagating with a finite speed, the flash from the assistant's lantern would have been observed by Galileo with a certain delay. The result of this experiment was, however, completely negative, and we know now very well why. Light propagates so fast that the expected delay in Galileo's experiment must have been about one hundred-thousandth of a second, which is quite unnoticeable to human senses.

The first successful zmeasurement of the velocity of light was carried out in 1675 by the German astronomer, Roemer, who replaced Galileo's assistant by the moons of the planet Jupiter, thus increasing the distance to be covered by light by a factor of hundreds of millions. Roemer's method is illustrated which shows the orbits of the earth, Jupiter, and one of its moons. Moving around the planet, the moons are periodically eclipsed as they enter the broad cone of shadow cast by Jupiter. Studying these eclipses, Roemer noticed that sometimes they took place as much as eight minutes ahead of schedule and sometimes with a delay of eight minutes. He also noticed that the eclipses were early when the earth and Jupiter were on the same side of the sun (1st position) and delayed in the opposite case (2nd position).

Ascribing correctly the observed irregularities to the difference of time taken by light to cover the changing distance between the earth and Jupiter, Roemer calculated that light must be propagating through space at a speed of about 300,000 kilometers per second (3×10^{10} cm/sec). The first laboratory measurement of the speed of light was carried out in 1849 by the French physicist, H.L. Fizeau (1819-1896), whose apparatus is shown in Fig. 9. It consists essentially of a pair of cogwheels set at the opposite ends of a long axle. The wheels were positioned in such a way that the cogs of one were opposite the intercog openings of the other, so that a light beam from the source could not be seen by the eye no matter what the position of the wheels.

However, if the wheels were set in fast rotation and the speed of this rotation was such that the wheels moved by half the distance between the neighbouring cogs during time taken by light to propagate from one wheel to the other, was exported to pass through without being stopped. In order to observe the effect at the speed of a few thousand revolutions per minute, which was about the maximum that Fizeau could achieve, he had to lengthen the path of the light beam by using four mirrors as shown in Fig. 9. This direct laboratory measurement gave a value for the speed of light that stood in reasonably good agreement with that obtained by Roemer's astronomical method.

If, for example, Fizeau used wheels with 100 teeth spinning at 3,000 rev/min, we can compute the necessary lengthening of the path of the light rays. While the light travels to the distant mirrors and back again, the wheels must have turned through the angle between a tooth and a space, in 1/200 of a revolution. At 3,000 rev/min this would require:

We know that the speed of light is almost exactly 3×10^{10} cm/sec, so in this time interval of 10^{-4} sec it would have had to travel 3×10^{6}cm or 30 km. Thus, I, the distance to the far mirrors, would have to be about 15 km or approximately 9 min. Later more accurate determinations of c, as the speed of light is universally called, were made by the American physicist, A. A. Michelson, by a similar scheme that used a rotating mirror instead of a toothed cogwheel.

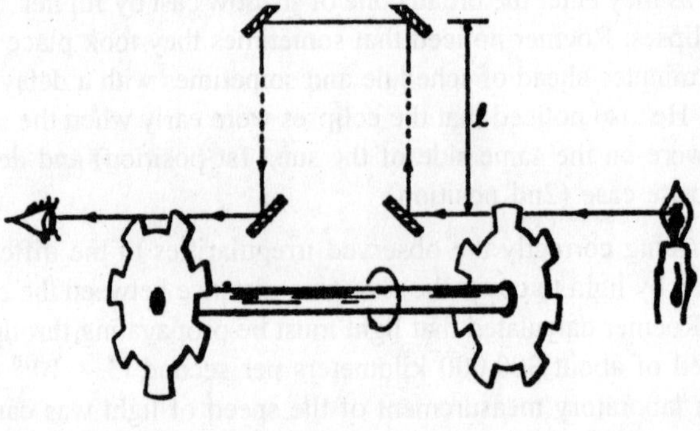

Fig. 9 : Fizeau's method for measuring the velocity of light.

$$\frac{1}{3{,}000 \times 200} = \frac{1}{600{,}000} \text{ min, or } 10^{-4} \text{ sec}$$

More recently, a polarization phenomenon has been employed to serve as a more accurately timed shutter than rotating wheels or mirrors. Certain substances, such as glass and many organic liquids, become doubly refracting when they are in a strong electric field. The Kerr cell is a glass container, generally filled with nitrobenzene, that is placed where the doubly refracting material. The orientation of polarizer and analyzer are adjusted so that no light normally passes through. But when an electronic device creates an electric field in the cell, the nitrobenzene becomes doubly refracting, and a flash of light can pass through the analyzer. Modem electronics can time these flashes to an accuracy of less than 10^{-6} sec, and very accurate values of c can be determined without having to use long light paths.

Work is still constantly being done by new and increasingly refined methods, and we are now fairly sure that the velocity of light in a vacuum is not far from 2.99776×10^{10}cm/sec.

RAMAN EFFECT

In 1928, Sir C.V. Raman discovered that when a beam of monochromatic light was passed through an organic liquid such as benzene, the scattered light was found to contain a strong line of frequency v_o equal to the frequency of the incident light and a few weak lines on either side of the incident line $v' = v_o \pm v_M$. This is referred to as *Raman Scattering*. The scattering of light with change of frequency is known as *Raman effect*. The spectrum formed due to Raman effect is called *Raman spectrum* and the spectral lines obtained are called *Raman lines*. In order to observe Raman effect, the incident light should be monochromatic and very intense. Raman scattering is always accompanied by Rayleigh scattering. Raman lines at frequencies less than that of the incident frequency ($v_0 - v_M$) are known as Stokes lines and those with frequencies greater than that of the incident frequency ($v_0 + v_M$) are known as *anti-Stokes* lines. Experiments show that Stokes lines are far more intense than anti-Stokes lines. Overall, however, the total radiation scattered at any but the incident frequency is extremely small, and sensitive apparatus is needed for its detection and study.

If the frequency of the incident light is varied, the weak lines are once again observed on either side of the Rayleigh line with the same difference in frequency. It is evident that the frequency difference v' between the incident and the scattered light in Raman effect is determined by the nature of the scattering molecules and is independent of the frequency of incident light.

CLASSICAL EXPLANATION

When a sample of molecules is subjected to a beam of light of frequency v, the induced dipole undergoes oscillations of frequency v;

$$\mu = \alpha E = \alpha E_0 \sin 2\pi vt \qquad ...(1)$$

If in addition, the molecule undergoes some internal motion, such as vibration or rotation, which changes the *polaribility* periodically, then the oscillating dipole will have superimposed upon it the vibrational or rotational oscillation. As an example, let us assume a vibration of frequency v_{vib}, which changes the polarizability. Then

$$\alpha = \alpha_0 + \beta \sin 2\pi v_{vib} t \qquad ...(2)$$

where α_0 is the equilibrium polarizability and p represents the rate of change of polarizability with the vibration. Using (3) into (52), we get

$$\mu = \alpha E = [\alpha_0 + \beta \sin 2\pi v_{vib}\, t]\, E_0 \sin 2\pi vt \qquad ...(3)$$

Expanding the above equation and using the trigonometric relation, we get

$$\mu = \alpha_0 E_0 \sin 2\pi vt + 1/2\, \beta E_0 [\cos 2\pi (v - v_{vib})\, t - \cos 2\pi (v + v_{vig})\, t\} \qquad ...(4)$$

It means that the oscillating dipole has frequency components $v \pm v_{vib}$, asrwell the exciting frequency v. These additional frequencies are responsible for producing the Raman lines. Thus, for Raman scattering, *a molecular rotation or vibration must cause some change in a component of molecular polarizability.*

Calcite Crystal

The most common example of uniaxial crystals is calcite. Calcite crystals are rhombohedral, as seen from Fig. 10. It is bounded by six faces, each of which is a parallelogram with angles equal to 102° and 78°. The angles between the edges meeting at each of the two diametrically opposite corners A and H are obtuse (larger than 90°). Hence, these two comers are called *blunt comers* of the crystal.

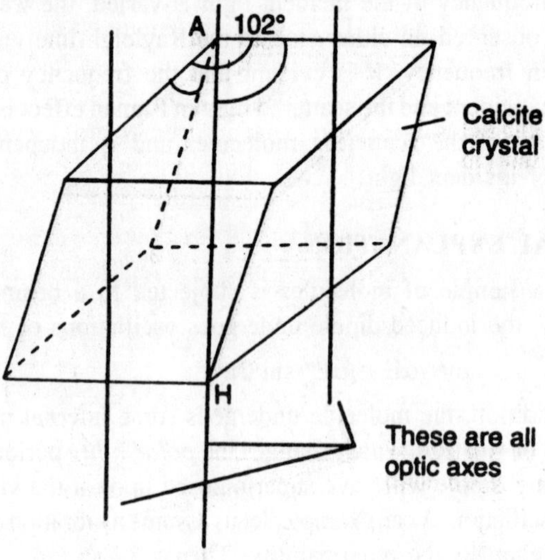

Fig. 10

OPTIC AXIS

A line bisecting any of the blunt comers is the optic axis. In fact any line parallel to this line is also an optic axis. Thus, the optic axis is a direction and not a specific line in the crystal. It is to be noted that the optic axis is not obtained by joining the two blunt comers. Only in a special case, when the three edges of the crystal are equal, the line joining the two blunt comers A and H coincide with the crystallographic axis of the crystal and it gives the direction of the optic axis. The optic axis is actually the axis of symmetry of the crystal. A ray of light propagating along optic axis does not suffer double refraction, because the structure of the crystal is symmetric about that direction.

The optic axis is the direction in a uniaxial crystai along which the e-ray and the o-ray travel with the *same* speed and consequently double refraction does not take place along this direction. The corresponding refractive index is the refractive index for ordinary light, say μ_o.

Principal Section

A plane containing the optic axis and perpendicular to a pair of opposite faces of the crystal is called the *principal section* of the crystal for that pair of faces. Thus, there are three principal sections passing through any point within the crystal, one corresponding to each pair of opposite faces. A principal section always cuts surfaces of calcite crystal in a parallelogram having angles 71° and 109°. The lines parallel to AB represent the end-views of other principal sections parallel to AB with in the crystal. Defining principal section is not enough to understand the directions of vibrations for the o-ray and e-rays. Hence, two more planes are defined as principal plane for the o-ray and the principal plane for the e-ray. The plane containing the optic axis and the o-ray is called the principal plane of the o-ray and the plane containing the optic axis and the e-ray is called the *principal plane of the e-ray.* The directions of vibrations in the o-ray and e-ray can be understood with reference to these planes.

MILLER INDICES

A crystal lattice may be considered as an aggregate of a set of parallel equidistant planes passing through the lattice point. These are called lattice planes. For a given lattice these sets of planes can be chosen in an umber of different ways; the spacing between the successive planes accordingly vanes, as also the density of lattice points per unit area in each plane.

A crystal can be easily split or *cleaved* along these lattice planes, particularly the planes of high density of lattice points. Hence lattice planes are also called *cleavage planes.*

The most important cleavage planes are those defined by (a, b) (b, c) and (c, a) But numerous other lattice planes may be defined. The method of specifying them is a follows:

(i) Let the intercepts by the given plane on the three crystal axes be in the ratio pa : qb : rc where a, b, c are the corresponding unit vector lengths, and p, q, r are integers.

(ii) Take the reciprocals of p, q, r.

FEE LATTICES PLANES

In a simple cubic lattice the largest spacings d are given by the smallest values of $h^2 + k^2 + l^2$. Thus largest three spacings are as follows:

planes	spacing	ratio
{100}	a	1.00
{110}	$\frac{a}{\sqrt{2}}$	0.71
{111}	$\frac{a}{\sqrt{3}}$	0.58

In the fcc lattice additional lattice points appear at the locations {a/2, all, a}. Now the {100} and {110} planes do not cover the face-centred points and we have to draw additional planes half-way through in each case.

However, the (111} planes take care of the additional points without introducing extra planes.

Mathematically the decision follows from eq. (2), which for cubic lattice becomes

$$hx + ky + lz = na \quad ...(1)$$

The {a/2, a/2, 0} points do not satisfy this for the {100} and {110} planes; they do for the {111} planes.

Some authors specify the half-way planes by using Miller indices {200} and {220}, which is a modification of step (iii) to define h, k, *l*.

This conveniently gives the spacings of Miller planes in fcc lattice.. The lowest three $h^2 + k^2 + l^2$ now correspond to the planes as listed below:

planes	spacing	ratio
{111}	$\frac{a}{\sqrt{3}}$	1.00
{200}	$\frac{a}{\sqrt{2}}$	0.87
{220}	$\frac{a}{2\sqrt{2}}$	0.61

For the bcc lattice, one may see similarly that additional planes arise half-way in the cases of {100} and {111} orientations, but not in the case of {110}. Hence the first two planes become {200} and {222} and hence planes {110} have the largest spacing. The three largest spacings are as follows:

planes {110}	spacing $\frac{a}{\sqrt{2}}$	ratio 1.00
{200}	$\frac{a}{2}$	0.71
{222}	$\frac{a}{2\sqrt{3}}$	0.41

SOLVED EXAMPLES

Example 1:

The smallest angle for strong reflection of a beam of neutrons with a family of crystallographic planes of spacing 3.84A is 30°. Calculate the wavelength of the neutron beam and also the speed of neutrons.

Solution:

From the Bragg equation $2d \sin\theta = n\lambda$.

$$2 \times 3.84 \times 10^{-8} \times 0.500 = 1\lambda$$

We put n = 1 since the angle is the smallest,

$$\therefore \qquad \lambda = 3.84 \times 10^{-8} \text{ cm}$$

Now, by de Broglie relation

$$\lambda = \frac{h}{p} = \frac{h}{mv}$$

$$\therefore \qquad v = \frac{h}{\lambda m} = \frac{6.63\times10^{-17}}{3.84\times10^{-8}\times1.67\times10^{-24}}$$

$$= 1.03 \times 10^{5} \text{ cm/sec.}$$

Example 2:

In a tetragonal lattice a = b = 2.42A, c = 1.74A. Deduce the lattice spacings between (1 0 1) planes.

Solution:

We have

$$d_{101} = \left(\frac{1^2}{(2.42)} + \frac{0^2}{(2.42)^2} + \frac{1^2}{(1.74)^2}\right)^{-1/2}$$

$$= 1.41\text{A (check it).}$$

Example 3:

For a given metal surface the limiting wavelength for photoelectric emission if 5800Å. Deduce the work function, maximum kinetic energy of electrons ejected by light of λ *= 4500A.*

Solution:

$$\text{Photon energy } h\nu = \frac{6.6\times10^{-34}\times3\times10^{10}}{4500\times10^{-8}}$$

$$= 4.40 \times 10^{-19}\text{J}$$

$$\text{Work function} = h\nu_o = \frac{6.6\times10^{-34}\times3\times10^{10}}{5800\times10^{-8}}$$

$$= 3.41 \times 10^{-19}\text{J}$$

The difference gives Ek = 0.99×10^{-19}J

Example 4:

Light of wavelength 5461A and intensity 0.20 $\times$ 10^{-5} *watt/cm*^{2} *falls on a photoelectric cell of cathode area 2.5 cm*^{2}*. The efficiency is 15%. Compute the saturation photocurrent.*

Solution:

Photons received *per second*

$$n = \frac{0.20\times10^{-5}\times2.5}{6.6\times10^{-34}\times3\times10^{8}\times1/(5461\times10^{-10})} = 1.37 \times 10^{13}\text{s}^{-1}$$

Electrons ejected per second are 15% of this; and each electron carries a charge 1.6×10^{-19} coulomb. Hence the saturation current is

$$i = 1.37 \times 10^{13} \times 0.15 \times 1.6 \times 10^{-19} \text{ amp}$$

$$= 0.33 \text{ microamp.}$$

Example 5:

In an experiment on Compton scattering the wavelength of incident radiation is 1.872 A. Calculate the wavelength of scattered radiation at θ *= 30°. Also calculate the velocity of the corresponding recoil electron.*

Solution:

By simple substitution for $\theta = 30°$

$$\Delta\lambda = .0242\ (1\text{-}8660) = .00324A^{\circ};$$

$$\lambda' = 1.872 + .003 = 1.875.$$

The energy difference of the incident and scattered photons gives the KE of the ejected electron. Hence

$$\frac{1}{2}mv^2 = h\ (v - v') = \frac{hc}{\lambda^2}(\Delta\lambda)$$

$$= \frac{6.6\times10^{-27}\times3\times10^{10}}{(1.872\times10^{-8})^2} \times (.00324 \times 10^{-8})\ \text{erg}$$

Substituting for m and solving we get

$$v = 2 \times 10^8\ \text{cm/sec.}$$

Now, from conservation of momentum, remembering that $\sqrt{}/v = 1$.

$$mv \cos\phi = \frac{hv}{c} - \frac{hv'}{c} \times .866$$

$$= \frac{hv'}{c} \times 0.134$$

$$mv \sin\phi = \frac{hv'}{c} x\ 0.500$$

$$\therefore \qquad \tan\phi = \frac{0.5}{0.134} = 3.62 \Rightarrow \phi = 75^{\circ}.$$

Example 6:

One sectorial series in an element is given by

$$v_n = \frac{109756}{(2-0.083)^2} - \frac{109756}{(n-0.643)^2},\ n = 3,\ 4,\ 5,...$$

Write down two more spectral series using Ritz principle, and evaluate v for the fist member of each series.

Solution:

We predict one series by changing the integer in the first term denominator to 3. Thus

$$v_n' = \frac{109756}{(3-0.083)^2} - \frac{109756}{(n-0.0643)^2},\ n = 4,\ 5,\ 6,...$$

We predict one more series by using the first running term of the given series as the fixed term, and changing the fixed term to running terms. Then

$$v_n'' = \frac{109756}{(3-0.643)^2} - \frac{109756}{(n-0.083)^2} \; n = 3, 4, 5,...$$

For the new series thus predicted the v values of the first term may be now calculated

$$v' = 3.17 \times 10^3 \text{ cm}^{-1};$$

$$v'' = 6.85 \times 10^3 \text{ cm}^{-1}.$$

Example 7:

About 0.1% of electrical energy supplied to a laboratory mercury vapour lamp of 80 watt is converted into UV light of wavelength 2500Å. Calculate the number of UV photons emitted per second by the lamp.

Solution:

Number of UV photons emitted per second

$$= \frac{\text{Energy conveted int o UV light}}{\text{Energy carried by one UV photon}}$$

$$= \frac{80J/s \times (0.1/100) \times (6.24 \times 10^{18} eV/J)}{12400eV/2500}$$

$$= 10^{17} \text{ photons/s.}$$

Example 8:

Deduce the average energy of Planck's oscillator corresponding to v values such that hv/kT = 0.1, 1.0, 3.0, 10.0.

Solution:

We have

$$\bar{\varepsilon} = kT \frac{x}{e^x - 1}, \text{ where } x = \frac{hv}{kT}$$

For the four given vales of x, tables show

$$\varepsilon^{x-1} = 0.105, 1.718, 19.1, \text{ and } 2.2 \times 10^4$$

Hence the value of ε is given by

$$\bar{\varepsilon} = kT\left(\frac{.1}{.105}, \frac{1}{1.718}, \frac{3}{19.1}, \frac{10}{2.2 \times 10^4}\right)$$

$$= kT\ (1{,}0.58,\ 0.16,\ 0.00045)$$

Thus for hv, < < kT we have ε = kT while for hv > > kT, ε tends to zero.

Example 9:

A surface is placed 20 cm from a 50W source of λ = 6x 10^{-5} cm. Estimate (i) the average time interval between the arrival of two photons on the same atom (sectional area – 10^{-16}cm^2, and (ii) the mean time lag between irradiation and ejection of the first electron from the surface (area 2 cm^2, efficiency 40%).

Solution:

Energy flux = $50/4\pi\ (0.20)^2 = 100\ \text{Jm}^{-2}\text{s}^{-1}$

Photon energy $= hc/\lambda = 6.6 \times 10^{-34} \times 3 \times 10^{+8}/6 \times 10^{-7}$

$= 3 \times 10^{-19}$ J/photon

$\therefore$ Photon flux at 20 cm $= 100/(3 \times 10^{-19})$

$= 3 \times 10^{20}$ photon $\text{m}^{-2}\text{s}^{-1}$

Now, (i) photon flux on a given atom = 3 photon s^{-1}

λ mean Δt = 0.3s

(ii) ejection rate from 2 cm^2 area $= 3 \times 10^{20} \times 2 \times 10^{-4} \times .4$

$= 2.4 \times 10^{16}$ electrons^{-1}

$\therefore$ mean $\Delta t = 4 \times 10^{-15}$s

(*Note*: One conclusion is that the chances of two photons acting together in ejecting an electron are negligibly small. Another is that the time lag between irradiation and start of emission is too small to be observable.)

Example 10:

X-rays of 0.5Å are scattered by free electrons in block of carbon through 90°. Find the velocity of recoil electrons.

Solution:

K.E. of the recoil electrons $= h\nu_i - h\nu_f$

$$= 12400\left(\frac{1}{\lambda_i} - \frac{1}{\lambda_f}\right)\text{eV}\ (1.602 \times 10^{-19}\ \text{J/e})$$

$$\lambda_f = \lambda_i + \frac{h}{m_0 c}(1 - \cos\theta)$$

$$= (0.5 + 0.02426)\text{Å} = 0.5243\text{Å}$$

$$\therefore \quad \text{K.E.} = 12400\left(\frac{1}{0.5}-\frac{1}{0.5243}\right)(1.602 \times 10^{-19})\ \text{J}$$

$$\therefore \quad u = \left[\frac{2(\text{K.E.})}{m}\right]^{1/2} = \left[\frac{2(1.84\times10^{-16}\,\text{J})}{9.11\times10^{-31}\,\text{Kg.}}\right]^{1/2}$$

$$= \left\{4.04\times10^{14}\,\frac{\text{kg.m}^2/\text{s}^2}{\text{kg}}\right\}^{1/2}$$

$$\therefore \quad u = 2 \times 10^7 \text{ m/s.}$$

Example 11:

Calculate de Broglie wavelength of an electron moving with velocity 10^7 m/s.

Solution:

$$\lambda = \frac{h}{mu} = \frac{6.626\times10^{-34}\,\text{J.s}}{9.11\times10^{-31}\,\text{kg}\times10^{7}\,\text{m/s}}$$

$$= \frac{6.626\times10^{-34}}{9.11\times10^{-24}}\frac{\text{kgm}^2/\text{s}^2.(\text{s}^2)}{\text{kg.m}} = 7.28 \times 10^{-11}\ \text{m}$$

$$\therefore \quad \lambda = 0.72\text{Å}$$

Example 12:

In a Bragg spectrometer a photographic film is set perpendicular to the incident beam and at a distance 25.0 cm from the axis of the NaCl crystal. A line is observed at a distance 8.472 cm from the direct beam. Treating this as a second order reflections from {110} face, deduce the wavelength. [a = 5.63A].

Solution:

$$\tan 2\theta = \frac{8.472}{25.0},$$

$$\therefore \quad \theta = 9°22' \text{ (check it)}$$

Now, for {110} reflections, $d = a\sqrt{2}$

$$\therefore \quad \lambda = \frac{2d\sin\theta}{n} = \frac{2\times5.63\times\sin 9°22'}{\sqrt{2}\times2}$$

$$= 0.651\text{A.}$$

Example 13:

In a triclinic crystal[3] a cleavage plane makes intercepts 2.93, 4.47 and 2.35 units along the three crystallographic axes, the corresponding primitives being 3.05, 6.99 and 4.90 A. Deduce the Miller indices of the cleavage plane.

Solution:

$$p:q:r = \frac{2.93}{3.05}:\frac{4.47}{6.99}:\frac{2.35}{4.90}$$

$$= .961: .640: .480$$

$$= 6:4:3$$

$$\therefore \quad p^{-1} : q^{-1} : r^{-1} = \frac{1}{6}:\frac{1}{4}:\frac{1}{3} = 2:3:4$$

Hence the Miller indices of the plane are (234).

Example 14:

Compute the minimum uncertainty in the location of a mass of 2.0gm moving with a speed of 1.5 m/s and the minimum uncertainty in the location of an electron moving with a speed of 0.5 × 10⁸ m/s. Given that the uncertainty in the momentum p for both $\Delta p = 10^{-3}p$.

Solution:

$$p = mu = 2 \times 10^{-3}\text{ kg} \times 1.5\text{ m/s} = 3 \times 10^{-3}\text{ kg m/s.}$$

$$\therefore \quad \Delta p = 10^{-3}\,p = 3 \times 10^{-6}\text{ kg m/s}$$

$$\therefore \quad \Delta x = \frac{h}{2\pi\Delta p} = \frac{6.63\times10^{-34}\,\text{Js}}{2\pi\times3\times10^{-6}\,\text{kg.m/s}}$$

$$= 3.5 \times 10^{-19}\text{Å}$$

For an electron, $p = mu$

$$= (9.11 \times 10^{-31}\text{ kg})\,(0.5 \times 10^{8}\text{ m/s})$$

$$= 4.55 \times 10^{-25}\text{ kg. m/s}$$

$$\therefore \quad \Delta p = 10^{-3}\,p = 4.55 \times 10^{-26}\text{ kg. m/s}$$

$$\therefore \quad \Delta x = \frac{h}{2\pi\Delta p} = \frac{6.63\times10^{-34}\,\text{Js}}{2\pi\times4.55\times10^{-26}\,\text{kg.m/s}} = 23\text{Å}.$$

Example 15:

With Franck-Hertz type of experiment on sodium vapour the first spectral line to appear is the D-line, $\lambda = 5.89x\ 10^{-5}$ cm. Deduce the first excitation potential of sodium.

Solution:

If excitation potential is V volt, excitation energy is eV joule where $= 1.6 \times 10^{-19}$ coul. Equating this with hv,

$$1.6 \times 10^{-19} V = hv = \frac{hc}{\lambda} = \frac{6.62 \times 10^{-34} \times 3 \times 10^{8}}{5.89 \times 10^{-7}}$$

$$V = \frac{6.62 \times 3}{5.89 \times 1.6} = 2.1 \text{ volt.}$$

Example 16:

Sodium has first two excited states at 2.1 and 3.7 eV. In a Franck-Hertz experiment electrons of energy 4.7 eV are fired in sodium gas. Deduce the possible energy values of the electrons received at the collection.

Solution:

In collisions the electrons may lose energy 0, or 2.1 eV, or 3.7 eV or 2×2.1 eV or $(2.1 + 3.7)$ eV, or (2×3.7) eV... Out of these the values below 4.7 eV are 0, 2.1, 3.7, and 4.2 eV. Hence the possible energies of the received electrons are 4.7 eV *minus* these values. Thus

$$E = 4.7, 2.6, 1.0, 0.5 \text{ eV.}$$

Example 17:

The density of sodium chloride is 2.18 g/cc. Given that the crystal has fcc space lattice deduce the lattice constant a. (Avogadro number $= 6.02 \times 10^{23}$ per g-mole).

Solution:

Volume of unit cell is a^3 cc and mass in each unit cell is therefore $a^3 \times 2.18$ g.

Now molecular weight of NaCl is 58.5, and in *fcc* lattice there are 4 molecules per unit cell. Hence mass in each unit cells is

$$(4 \times 58.5), (6.02 \times 10^{23})\text{g}$$

Equating the two expressions.

$$a^3 \times 2.18 = \frac{4 \times 58.5}{6.02 \times 10^{23}} \Rightarrow a = 5.63 \times 10^{-8} \text{ cm}.$$

Example 18:

Calculate the glancing angle on the cube face {100} of a rocksalt crystal (a = 2.814A) corresponding to second order reflection for X-rays of λ = 0.710A.

Solution:

For {100} planes, the spacing d is equal to a.

$$2 \times 2.8114 \times 10^{-8} \sin\theta = 2 \times 0.710 \times 10^{-8}$$

$$\theta = \sin^{-1}\left[\frac{0.710}{2.814}\right] = 14°22'.$$

Example 19:

Packing fraction in crystals is defined as the ratio of volume of the atoms in a unit cell and the volume of unit cell. Compute it for (i) sc, (ii) fcc, and (iii) hcp structure of elemental solids, treating the atoms as spherical.

Solution:

If atomic diameter is d, the volumes occupied by the atoms in a unit cell in the three cases are:

(i) $[\pi d^3/6] \times 1$ (ii) $[\pi d^3/6] \times 4$ (iii) $[\pi d^3/6] \times 2$

Unit cell parameters for the cases are:

(i) $a = d$, (ii) $a = d\sqrt{2}$,

(iii) $a = b = d$, $c = 2d\sqrt{2/3}$, $\lambda = 120°$

Hence the unit cell volumes are:

(i) d^3, (ii) $2\sqrt{2}\,d^3$, (iii) $\sqrt{2}\,d^3$

The required packing fractions therefore are:

(i) $\pi/6$, (ii) $\pi/3\sqrt{2}$, (iii) $\pi/3\sqrt{2}$.

Example 20:

Show that the (1 1 0) planes do not cover all the face centre points in a fcc lattice.

Solution:

Face centered points are represented by {ra/2, sa/2, ta} where r and s are odd integers and t is any integer. The curly bracket here means

$$\left(\frac{ra}{2}, \frac{sa}{2}, ta\right), \left(ta, \frac{ra}{2}, \frac{sa}{2}\right)$$

and $$\left(\frac{sa}{2}, ta, \frac{ra}{2}\right)$$

(1 1 0) planes refer to h = 1, k = 1, l = 0. Substitution for the three sets of face-centered points,

$$\frac{ra}{2} + \frac{sa}{2} = na,\ ta + \frac{ra}{2} = na,$$

and $\frac{sa}{2} + ta = na$

Only the first of these holds true (since r and s ae odd integers), which mans that the other sets of the face-centered points are *not* covered by the (1 1 0) planes.

Example 21:

X-rays of λ = 0.741A are incident on a powder specimen. A plane film placed beyond the powder at distance 20.0 cm shows the smallest concentric rings of radii 7.22, 10.74, 17.06 and 32.50 cm. Interpret the observations regarding crystal structure.

Solution:

We deduce in succession tan 2θ, θ and sin θ for the four cases

$$\sin\theta = 0.1723,\ 0.2439,\ 0.3458,\ 0.4879$$

We note: that the last two value fo sin θ are very nearly two times the first two values. Hence it is *reasonable* to take the first two rings as the first order (n = 1) images and the last two as the second – order (n = 2) images.

From Bragg equation the two d values are related by

$$\frac{d_1}{d_2} = \frac{\sin\theta_2}{\sin\theta_1} = \frac{.2439}{.1723} = 1.416$$

This is close to $\sqrt{2}$ and we therefore conclude that $d_1 = d_2\sqrt{2}$.

In the cubic system the crystal could have sc or bcc; arrangement, but not fcc.

From the given λ, we get

$$d_1 = \frac{0.741}{2\times.1723} = 2.15A;$$

$$d_2 = \frac{7.31}{2\times.2439} = 1.52A.$$

Example 22:

Deduce the primitive vectors and primitive cell volume for a bcc lattice.

Solution:

The dotted lines show that cubic unit cell, with the body centre at O. The vectors of the primitive cell are lines from O to coroners A, B, C, (chose *any* one corner of the cube and go over across face-diagonals to get the other two corners; all choices are equivalent).

The co-ordinates required are

$$O\left(\frac{1}{2},\frac{1}{2},\frac{1}{2}\right)\ A\ (1,\ 0,\ 0)\ B\ (1,\ 1,\ 1)\ C\ (0,\ 0,\ 1)$$

So the vectors are (in units of a)

$$OA = a' = \left(\frac{1}{2}i-\frac{1}{2}j-\frac{1}{2}k\right),$$

$$OB = b' = \left(\frac{1}{2}i+\frac{1}{2}j+\frac{1}{2}k\right)$$

$$OC = c' = \left(-\frac{1}{2}i-\frac{1}{2}j+\frac{1}{2}k\right)$$

These lead to $a' = b' = c' = \frac{\sqrt{3}}{2}$ a.

$$\cos\lambda = \frac{a', b'}{a'b'} = \frac{\frac{1}{4}-\frac{1}{4}+\frac{1}{4}}{3/4} = -\frac{1}{3} \Rightarrow l = 109°28'$$

$$\text{volume} = c'.a' \times b' = \frac{1}{2}a^3 .$$

Example 23:

Alpha iron has a bcc lattice with lattice constant 2.86A°. Deduce the (i) co-ordination number, (ii) nearest neighbour distance, and (iii) number of atoms per unit cell.

Solution:

With any atom treated as reference atom, the nearest neighbours in a bcc lattice have positions

$$\left(\pm\frac{1}{2}, \pm\frac{1}{2}, \pm\frac{1}{2}\right) a$$

Thus the number of nearest neighbours is 2 × 2 × 2 = 8. This is the *co-ordination number*. The nearest neighbour distance is given by

$$d^2 = (a/2)^2 + (a/2)^2 + (a/2)^2 = a\sqrt{3/3}$$

In a unit cell in *bcc* lattice there are 8 atoms at the corners and 1 at the body centre. Since each corner atom is shared by 8 cells, there are (8/8) + 1 = 2 atoms per unit cell.

EXERCISES

1. Find the energy of neutron having de Broglie wavelength of 10^{-14}m. Given the rest mass of neutron is 1.6 × 10^{-27}kg.
2. Calculate de Broglie wavelength of an electron whose kinetic energy is 500 eV.
3. State the basic postulates of Planck's quantum theory. How did Einstein modify this theory in the light of experimental results?
4. The first member of Balmer series of hydrogen has a wavelength of 6536Å. Calculate the wavelength of its second member.
5. What is meant by work function of a material?
6. What is Compton effect? Show that Compton shift is independent of wavelength of incident x-rays.
7. Define crystal structure factor. Show that for CsCl the odd order maxima from (100) planes will be weak on this account.
8. A crystal involving equal number of atoms A and B has CsCl structure in the ordered system but in the disordered system A and B occupy the sites randomly. A and B have nearly equal atomic numbers, but have different magnetic moments. By what experiment will you decide the extent of ordering? Explain.
9. In a CsCI structure the positions of Cs atoms are main + $n_2$oj + $n_3$nk. Write down the positions of the Cl atoms, and the translation vector from any Cs atom to any Cl atom, or *vice-versa.*

10. The X-ray powder pattern of a simple solid of the cubic system is obtained on a plane film 20.0 cm away from the specimen. The radii of the first few rings are found to be 6.95, 10.28, 16.20, 22.2, and 29.7 cm. Deduce the Bragg angles and find whether the lattice is sc, bcc or fee.

11. Gold has atomic weight 197 and density 193 gm/cc. Deduce an order-of-magnitude value of spacing between the atoms in solid gold. (Given $N = 6.0 \times 10^{23}$ per gm mole).

12. Distinguish between cubic I and fee lattices. Compute the number of atoms of one kind in a unit cell in each case.

13. What is photoelectric effect? How does Einstein's equation explain the features of photoelectric effect?

14. Explain de Broglie's concept of matter waves.

15. Using the concept of matter waves, obtain the Bohr's condition for quantization of angular momentum.

16. What is Heisenberg Uncertainty principle? Explain how it is the out come of the wave description of a particle.

17. What is the physical significance of wave function 'ψ'?

18. A particle of mass 10^{-6} gm has a speed of 1m/s. If the speed is uncertain by 0.01% what is the minimum uncertainty in the position of the particle?

18. NaCl has fee lattice with a = 5.63Å. Deduce the spacings of {100} and {110} planes. Also compute the dimensions and shape of the primitive cell.

20. KCl has the same crystal structure as NaCl and both are ionic. But for reflections by {111} planes the odd order spectra are absent in KCl while they are only weak in NaCl. Discuss this. The atomic numbers of K, Na and Cl are 19,11 and 17 respectively.

21. In reflection of ordinary light the only condition required is $\theta' = \theta$, whereas in Bragg reflection an additional condition required is $2d \sin \theta = n\lambda$. Discuss this and also other differences between reflection of light from a surface and of X-rays from a lattice.

3
Polarization of Light

INTRODUCTION

Light coming from common light sources is unpolarized. However, they do not give any indication regarding the character of the waves. Whether the light waves are longitudinal or transverse, or whether the vibrations are linear or circular cannot be deduced from the above two phenomena, as all kinds of waves under suitable conditions exhibit interference and diffraction. In 1861 Arago and Fresnel showed that light waves vibrating in mutually perpendicular planes do not interfere. Interference and diffraction phenomena proved that light is a wave motion and enabled the determination of the wavelength. In 1817 Thomas Young explained the absence of interference by postulating that light waves are *transverse waves*.

About fifty years later. Maxwell developed electromagnetic theory and suggested that light waves are electromagnetic waves. As electromagnetic waves are transverse waves, it is obvious that light waves too are transverse waves. The concept of transverse nature leads to the concept of polarisation. The state of polarization cannot be detected by unaided human eye. An understanding of polarisation is essential for understanding the propagation of electromagnetic waves guided through wave-guides and optical fibres.

BREWSTER'S LAW

Sir David Brewster performed a series of experiments on the polarization of light by reflection at a number of surfaces. He found that the polarising angle depends upon the refractive index of the medium. In 1892, Brewster proved that *the tangent of the angle at which polarisation is obtained by reflection is numerically equal to the refractive index of the medium*. If θ_P is the angle and u. is the refractive index of the medium, then

$$\mu = \tan \theta_P \qquad ...(1)$$

This is known as Brewster's law.

If natural light is incident on a smooth surface at the polarizing angle, it is reflected along BC and refracted along BD. Brewster found that the maximum polarisation of reflected ray occurs when it is at right angles to the refracted ray. It means that $\theta_P + r = 90°$.

$$\therefore \qquad r = 90° - \theta_P \qquad ...(2)$$

According to Snell's law,

$$\frac{\sin \theta_P}{\sin r} = \frac{\mu_2}{\mu_1} \qquad ...(3)$$

where, μ_2 is the absolute refractive index of reflecting surface and μ_1 is the refractive index of the surrounding medium. It follows from eqn. (2) and eqn. (3) that,

$$\frac{\sin \theta_P}{\sin (90° - \theta_P)} = \frac{\mu_2}{\mu_1}$$

or

$$\frac{\sin \theta_P}{\sin \theta_P} = \frac{\mu_2}{\mu_1}$$

$$\therefore \qquad \tan \theta_P = \frac{\mu_2}{\mu_1} \qquad ...(4)$$

Eqn. (4) shows that the polarising angle depends on the refractive index of the reflecting surface. The polarising angle θ_P is also known as *Brewster angle* and denoted as θ_B. Light reflected from any angle other than Brewster angle is partially polarised.

APPLICATION OF BREWSTER'S LAW

(i) Brewster's law can be used to determine the refractive indices of opaque materials.

(ii) It helps us in calculating the polarizing angle necessary for total polarization of reflected light for any material if its refractive index is known. However, the law is not applicable for metallic surfaces.

(iii) Two windows known as *Brewster windows* are used in gas lasers. They are arranged at Brewster angle at the two ends of the laser tube. Every time light passes through the windows on its way to the reflecting mirrors of the optical resonator, s-component is removed. At the end, light emerging from the laser consists of linearly polarised light of p-component.

(iv) Another application utilises the Brewster angle for transmitting a light beam into or out of an optical fibre without reflection losses.

We have seen in an earlier chapter that there are two possible ways of wave propagation:

(1) Longitudinal waves in which the motion of individual particles is along the line of propagation, and

(2) Transverse waves in which the motion is perpendicular to the line of propagation. Which kind of wave motion do we have in the case of light?

The important difference between the longitudinal and transverse waves is that the latter can be "polarized." To understand this important notion let us look at a wave in the direction of its propagation as shown in Fig. 1. In the case of longitudinal waves (a), the motion of the particles takes place perpendicular to the surface of the paper and will not be noticeable from the direction we are observing it. In the case of transverse waves (b) and (c), the motion of the particles is in the plane of the paper and easily observable in that projection. We call the transverse wave "natural" or "nonpolarized" if the motion of the particles takes place in all possible directions (b); if the motion is only in one direction (c), the wave is "polarized."

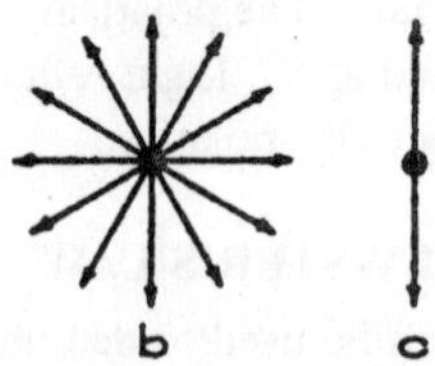

Fig. 1 : Cross-section of a wave in the direction of its propagation.

The notion of polarization can be clarified by the analogy. Suppose we have a sieve, made of a set of parallel wires without c oss wires, and drop matches on it in such a way that the falling matches remain horizontal but may have different orientations in the horizontal plane. It is clear that only those matches that are parallel to the sieve wires will pass through.

If under this "match polarizer" we place a similar "match analyzer," the "beam of matches" will pass through only if the wires in the lower sieve run parallel to those of the upper sieve. We can extend this analogy to the case of longitudinal waves by dropping the matches in a vertical position. In this case, all the matches will pass through, regardless of the relative positions of the two sieves.

To our eyes, the plane of vibration of light waves makes no difference—light that has been polarized so the vibration is all in a single plane Fig. 1(c) looks exactly the same as ordinary light that consists of myriads of photons whose planes of vibration are aligned in every possible direction Fig. 1(b). Much of the light we see is at least partially polluted, because whenever light is reflected from a smooth *non-metallic* surface the waves with vibrations parallel to the surface are reflected more intensely than those with vibrations perpendicular to the surface.

The light from the blue sky is also partially polarized. The, white light from the sun, in passing through the molecules of the air, has its blue component scattered, or reflected, from the molecules more than are the longer red waves, and this scattered blue light is partially polarized.

The eyes of bees can not only distinguish polarized light from non-polarized, but they can determine the plane of polarization. Bees apparently use this ability to tell directions and navigate back to the hive, and can do so even when the sun is covered by clouds as long as there is a patch of blue sky visible.

WAVES OF LIGHT ARE TRANSVERSE

For electromagnetic waves in the microwave region (λ + 1 cm) the transverse character can be demonstrated using a wire-grid polarizer. It has number of straight parallel wires (Fig. 2) of an electric conductor.

Let the plane of the grid bexy plane with length of wires along JC-direction, and let the waves travel along the z-direction. If one more wire-grid (not shown) is placed parallel to the first and is rotated in its own plane, then no waves pass if the wires in the two grids are mutually perpendicular, while waves pass if the wires in the two grids are parallel.

The explanation is simple. The x-oscillations of electric field E in the e.m. waves make the electrons in the wires oscillate and the energy is thus dissipated as heat They-osallations of E field are unable to do this (assuming wire diameter $<< \lambda$) and hence pass through.

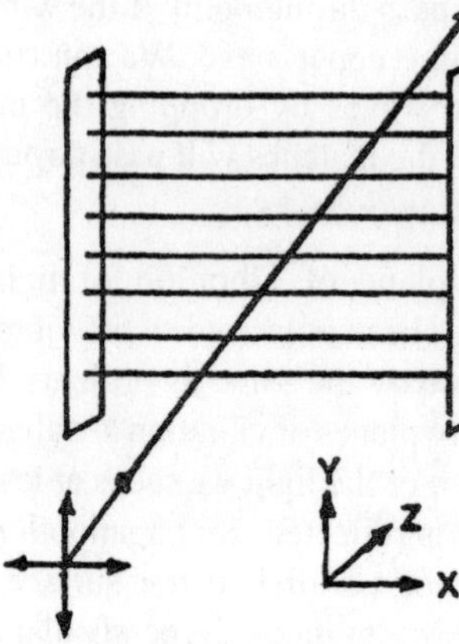

Fig. 2. : The wire-grid polarizer.

POLARISED LIGHT

The simplest type of an electromagnetic wave is a wave in which the direction of vibrations of electric vector E is strictly confined to a single direction in the plane perpendicular to the direction of propagation of the wave. Such a light is said to be *plane polarised light.* If the wave is coming towards the eye, the electric vector appears executing a linear vibration normal to the ray direction. Hence, a plane-polarised wave is also known as a linearly polarised wave. We therefore offer the following definitions:

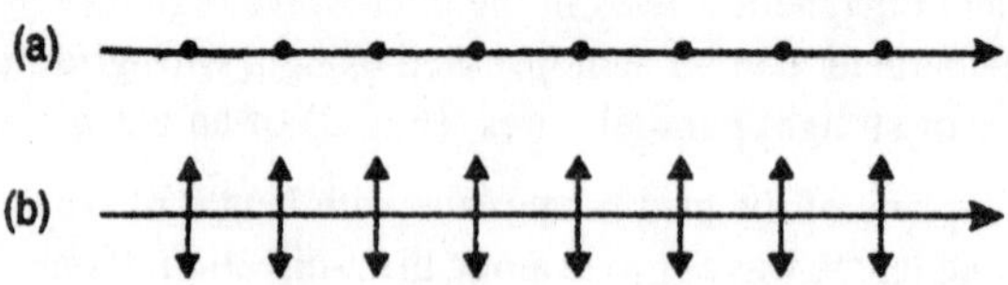

Fig. 3

A *plane-polarised* light wave is a wave in which the electric vector is everywhere confined to a single plane. A *linearly polarised* light wave is a wave in which the electric vector oscillates in a given, constant orientation. Now-a-days the adjective "linearly polarised" is more preferred and frequently used than "plane polarised". Ray diagrams, as shown in Fig. 3, represent the linearly polarised light.

Natural Light

An ordinary light source consists of a very large number of atomic emitters. Each atom radiates a plane polarised wave train for about 10^{-8}s. As time progresses the orientation of the plane of polarisation changes randomly and rapidly at the rate of 10^8 times per second. Our eyes cannot respond to such rapid changes. In the time that the eye responds to light, millions of wave trains are emitted by the source. Further, the frequencies of the wave trains are not exactly identical.

Thus, ordinary light comprises of a heterogeneous group of wave trains having different wavelengths and vibrating in different planes perpendicular to the direction in which the ray is propagating (Fig. 4a).

The random orientation of the planes of polarisation leads to equal probability of all directions about the propagation direction, as shown in (Fig. 4b). Therefore, in natural light the optical vector has no specific favoured orientation. *Light in which the planes of vibration are symmetrically distributed about the propagation direction of the wave is known as the unpolarized light.*

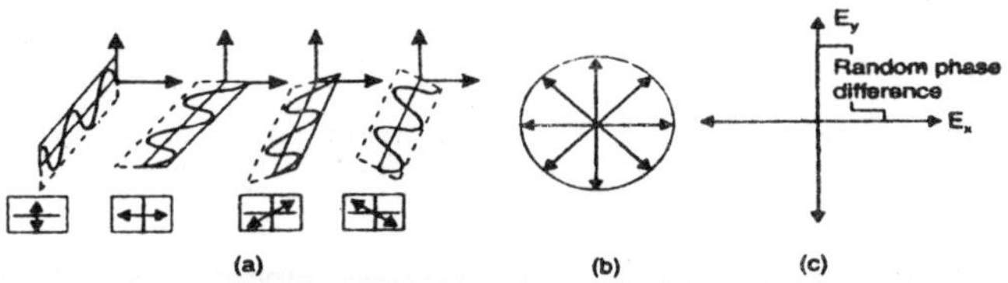

Fig. 4 : Unpolarized light (a) component waves having different planes of polarization (b) Schematic representation, (c) Representation of unpolarised light.

We may resolve each optical vector in unpolarized light into two rectangular components lying parallel and perpendicular to a chosen direction (Fig. 4c). Instead of showing a large number of optical vectors as in (Fig. 4b), we use only two vectors, as in (Fig. 4c), for representing unpolarized light.

Due to the random distribution of the optical vectors, the amplitude of the component vectors will be equal. Therefore, if the intensity of the incident unpolarized light is I_0 the intensity of each component will be $1/2\ I_0$. Thus, the intensity of unpolarized light is equally distributed between the two components.

DIRECTION OF THE WAVE

Waves are basically of two types, namely longitudinal waves and transverse waves. A wave in which particles of medium oscillate to and fro along the direction of wave propagation is called a *longitudinal wave*. Sound waves and waves produced on a spring are examples of longitudinal waves. The longitudinal waves consist of alternate compressions and rare factions. In these waves, particle displacement (vibration) and wave propagation occur in the *same* direction. Therefore, there does not exist any preferential direction in the wave.

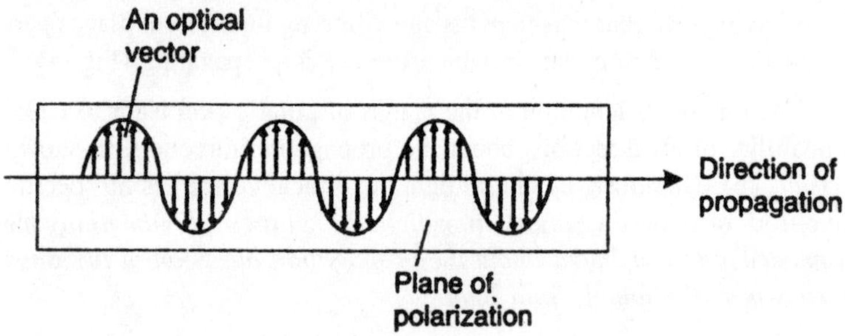

Fig. 5

A wave in which every particle of the medium oscillates *up* and *down* at right angles to the direction of wave propagation is called a transverse wave. Ripples on water surface and waves on a rope are examples of transverse waves. The wave propagation takes in the form of alternating crests and troughs. In a transverse wave, the direction of particle displacement occurs perpendicular to the wave propagation. Hence, the direction normal to the wave propagation is the *preferential direction* in a transverse wave.

The existence of a preferential direction leads to the characteristic phenomenon known as polarisation of transverse waves. Polarisation is not found with longitudinal waves as they do not possess a directional property. Polarisation is specific to transverse waves.

An electromagnetic wave is a transverse wave consisting of electric and magnetic fields vibrating perpendicular to each other and to the direction of propagation. The vibrating electric vector E and the direction of wave propagation form a plane. In olden days this plane was called the *plane of vibration*. However, it has become common now days to refer to it as the *plane of polarisation*.

We adopt the same notation and define the *plane of polarisation of the wave as the plane*, which *contains the optical vector E and the direction of propagation.*

BABINET COMPENSATOR

A *compensator* is an optical device whose function is to compensate a path difference. It is used in conjunction with a polarizer and analyzer combination to investigate elliptically polarized light. The compensator helps in determining the axis of the ellipse and the ratio of their lengths. Elliptically polarized light may be considered as the resultant of two coherent plane polarized waves occurring in mutually orthogonal directions and with an initial path difference of $\lambda/4$. When the elliptically polarized light passes through the device such that it introduces a further path difference of $\lambda/4$, the total path difference between the perpendicular waves becomes $\lambda/2$ and the vibrations recombine after emerging from the device to produce plane polarized light. From the analysis of this plane polarized light, the information regarding the incident elliptically polarized light can be obtained.

Construction

The Babinet compensator is made of two wedge-shaped quartz sections, ABC and ADC, having equal acute angles. The wedges are placed against each other such that they form a small rectangular block. One of the quartz wedges is fixed and the other can be displaced along their plane of contact with the help of a micrometer screw arrangement. Thus, the combination acts as a plate of variable thickness. The optic axis of the first section is parallel to its refracting edge AB and the optic axis of the second section is in a direction perpendicular to the edge. The two optic axes are perpendicular to each other and also perpendicular to the incident beam.

PRODUCTION OF POLARIZED LIGHT

Let plane polarized light be incident normally on the face AB of the compensator. It splits into e-ray and o-ray parallel and perpendicular to AB respectively. The e-ray travels slower than o-ray in the first section, since quartz is a positive uniaxial crystal.

When these rays enter the second section, the e-ray becomes o-ray since the optic axis in the second section is in a direction normal to that in the first prism. Similarly, o-ray becomes e-ray. Thus, the two rays exchange their velocities in passing from one section to the other section. The net effect is that one section cancels the effect of the other.

If d_1 is the thickness of the first section, and μ_e and μ_o are the refractive indices of quartz for e-and o-rays respectively, the path difference between the e- and o-rays in the first section will be

$$\Delta_1 = [\mu_e - \mu_o] \, d_1$$

As the principal planes of the two sections are at right angles, the e-and o-rays change their roles in going from the first section to the second section. The velocities of e-ray and o-ray interchange and if the thickness of the second section is d_2, then the path difference between the rays in the second prism will be

$$\Delta_2 = [\mu_e - \mu_o] \, d_2$$

As the compensator is thin, the separation of the rays is negligible. The net path difference between the two rays after emerging from the crystal will be

$$\Delta = \Delta_1 + \Delta_2$$

$$\Delta = (\mu_e - \mu_o) \, d_1 + (\mu_o - \mu_e) \, d_2$$

$$= (\mu_e - \mu_o) \, (d_1 - d_2) \qquad ...(1)$$

The net phase difference is

$$\delta = \frac{2p}{\lambda} (\mu_e - \mu_o) \, (d_1 - d_2) \qquad ...(2)$$

For a ray passing through the centre of the compensator where $d_1 = d_2$, the net path difference and hence the phase difference is zero. It means that the effect of one wedge is exactly cancelled by the other. This is true for all wavelengths and the incident wave is transmitted as such. Plane polarized light incident on the compensator will emerge as plane polarized light with its plane of vibration parallel to that of the incident light.

Any desired thickness difference $(d_1 - d_2)$ can be achieved at the centre of the compensator by moving the second section relative to the first section. Thus, any desired value of phase difference can be obtained between the e and o-rays. Therefore, the light emerging will be either plane or circular or elliptically polarized light, depending on the phase difference.

Thus, the compensator has the same effect as that of a wave plate of varying thickness. The advantage of compensator is that it can be arranged to suit any wavelength where as a quarter wave plate is designed to suit only one particular wavelength.

ANALYSIS OF ELLIPTICALLY POLARIZED LIGHT

Using the compensator, one can determine the characteristics of elliptically polarized light.

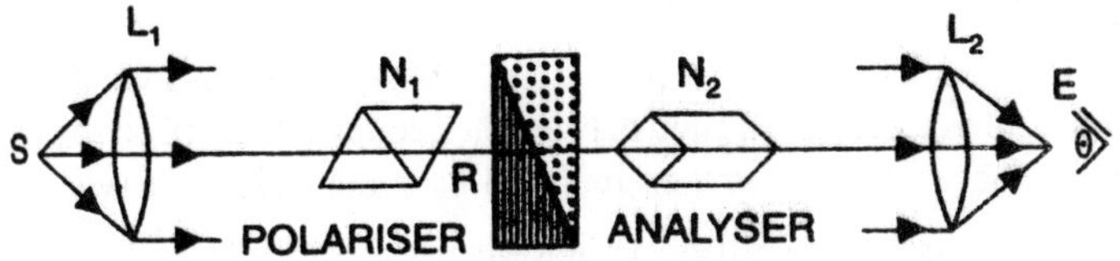

Fig. 6

Let the transmission axis of polarizer be oriented at 45° with respect to the optic axis of wedge ABC of the compensator. At midpoint R the light emergent from the compensator is plane polarized in the same plane as transmitted by N_1 and therefore it will be extinguished by the analyser N_2. Similarly, at distances from the midpoint for which the retardation is $1\lambda, 2\lambda, 3\lambda, \ldots .. w\lambda$, and the emergent light is plane polarized in the same plane as transmitted by N_1 and hence will be extinguished by the analyser. So the field of view is crossed by a series of equidistant parallel dark bands.

At positions between them, where the path difference corresponds to an odd multiple of $\frac{\lambda}{2}$, *i.e.*, $\frac{\lambda}{2}, \frac{3\lambda}{2}, \frac{5\lambda}{2}, \ldots .. \frac{(2m+1)\lambda}{2}$, the transmitted light is plane polarized. The analyser transmits the light completely and those regions will be bright. In all other cases, the emerging light is elliptically polarized with varying parameters of the ellipse, as shown in Fig. 6. If white light is used, the central band will be dark while others will be coloured.

By using white light source, the compensator is adjusted such that the central dark band is under cross wire and the micrometer reading is noted. The micrometer screw is turned through an angle such that the compensator introduces a phase difference of $\pi/2$ at cross wire.

Then elliptically polarized light is made to be incident on the compensator. The central dark band undergoes a shift with respect to the cross wire. The compensator is rotated through an angle a in its own plane until the central dark band is on the cross wire. The axes of the incident elliptically polarized light are parallel to the optic axes of the wedges of the compensator.

Phase Difference

The elliptical vibration can be regarded as made of two mutually perpendicular linear vibrations, which are having a phase difference, δ. δ can be determined as follows.

First the compensator is illuminated with white plane-polarised light and the micrometer is adjusted to bring the central dark band on the cross-wires. The white light is then replaced by elliptically polarised light.

The central band shifts to a point where the original phase difference δ between the two component vibrations of elliptical polarised light is exactly balanced by the phase difference introduced by the compensator. This phase difference is determined by rotating the screw until the central dark band is again on the cross-wires. If this rotation is ϕ, then

$$\frac{\delta}{2\pi} = \frac{\phi}{\alpha} \text{ or } \delta = \frac{2\pi\phi}{\alpha} \qquad ...(3)$$

Position of Axes

The position of the major and minor axes of the given elliptical vibration can be found as follows. The compensator is illuminated with white polarised light and the micrometer screw is adjusted to bring the central dark band on the cross-wires.

The screw is then turned through an angle $\alpha/2$ so that the compensator introduces a phase difference of 90°. The central dark band now is not on the cross wires. The elliptically polarised light is made incident on the compensator. Then the compensator is rotated in its own plane until the central dark band again comes on to the crosswires. The axes of the incident light are parallel to the optic axes of the wedges of the compensator.

Ratio of the Axes

Referring to the Fig. 7, OA and OB represent the optic axes of the two wedges of the compensator. OC is the direction of the principal section of the analyser. DD' is the direction of vibration of light emerging from the compensator at the crosswires. The tangent of the angle θ that the principal section of the analyser makes with the optic axes of the wedge gives the ratio of the axes. Thus,

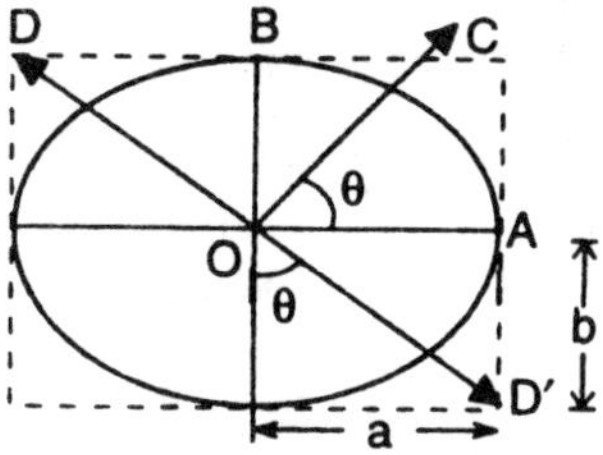

Fig. 7

$$\tan \theta = a/b$$

θ is determined by rotating the analyser until the bands disappear and the field becomes uniformly illuminated. The angle of rotation is θ.

Advantages : A quarter wave plate produces a fixed phase difference between o-ray and e-ray and can be used only for monochromatic light of one particular wavelength. In case of a compensator, the phase difference between the rays can be varied continuously and hence a compensator, made of a combination of wedges, can be used for light of any given wavelength.

ANALYTICAL TREATMENT

Let us assume that linearly polarised light is incident normally on a quartz plate cut perpendicular to the optic axis. Let the vibrations in the incident light be represented by

$$y = a \cos \omega t$$

On entering the crystal these vibrations are broken up into two equal and opposite circular vibrations which are represented by

$$x_1 = \frac{a}{2} \sin \omega t$$

and $y_1 = \frac{a}{2} \cos \omega t$ –clockwise circular polarisation

$$x_2 = \frac{a}{2} \sin \omega t$$

and $y_2 = \frac{a}{2} \cos \omega t$ –anti-clockwise circular polarisation

These circular components travel through the crystal with different velocities. When they emerge from the crystal, there is a phase difference ° between them. In case of quartz crystal, which is a right-handed optically active crystal, the clockwise component travels faster. The circular components are then represented by

$$x_1 = \frac{a}{2} \sin(\omega t + \delta)$$

and $y_1 = \frac{a}{2} \cos(\omega t + \delta)$ –clockwise component

$$x_2 = \frac{a}{2} \sin \omega t$$

and $y_2 = \frac{a}{2} \cos \omega t$ –anti-clockwise component

The resultant displacements along the two axes are

$$X = x_1 + x_2 = \frac{a}{2}\left[\sin(\omega t + \delta) - \sin \omega t\right]$$

$$= a \cos\left[\omega t + \frac{\delta}{2}\right] \sin \frac{\delta}{2} \quad ...(1)$$

and $Y = y_1 + y_2 = \frac{a}{2}\left[\cos(\omega t + \delta) + \cos \omega t\right]$

$$= a \cos\left[\omega t + \frac{\delta}{2}\right] \cos \frac{\delta}{2} \quad ...(2)$$

The resultant vibrations along the X-axis and Y-axis have the same phase. Therefore, the resultant vibration is plane polarised and it makes an angle $\delta/2$ with the original direction. Thus, the plane of polarisation is rotated through an angle $\delta/2$ on passing through the crystal.

Dividing eqn. (1) by eqn. (2, 3), we obtain

$$\frac{X}{Y} = \tan \frac{\delta}{2} \quad ...(3)$$

If μ_R and μ_L are the refractive indices of quartz in the direction of optic axis for clockwise and anticlockwise circularly polarised light and d is the thickness of the crystal plate, then the phase difference δ is given by

$$\delta = \frac{2\pi}{\lambda}(\mu_L - \mu_R)d \quad ...(4)$$

Hence the plane of polarisation is rotated through

$$\theta = \frac{\delta}{2} = \frac{\pi d}{\lambda}(\mu_L - \mu_R) \quad ...(5)$$

Experimental Verification of Fresnel's Theory

Fresnel showed that linearly polarised light on entering an optically active crystal is resolved into two circularly polarised vibrations. He arranged alternatively a number of negative and positive optically active prisms to form a rectangular block, as shown in Fig. 8. The optic axis is parallel to the base of each prism. When linearly polarised light is incident normally on the first crystal surface AB, the two component circular vibrations travel along the same direction but with different speeds. When the light beam passes through the first oblique surface BC, the R-component, which is faster in the first prism, becomes slower in the second prism.

The opposite is the situation for L-component. It means that the second prism BCD acts as a denser medium for the R-component and rarer medium for the L-component. As a result, the R-component bends towards the base in the second prism while L-component bends away from the base. Therefore, the two components are separated apart while they travel through prism BCD. Again at the boundary of the next prism, the speeds are interchanged and the R-component bends away from the base and the L-component bends towards the base. The net result is that the two beams are separated more and more as they pass through the successive prisms. When they ultimately emerge out, they are widely separated.

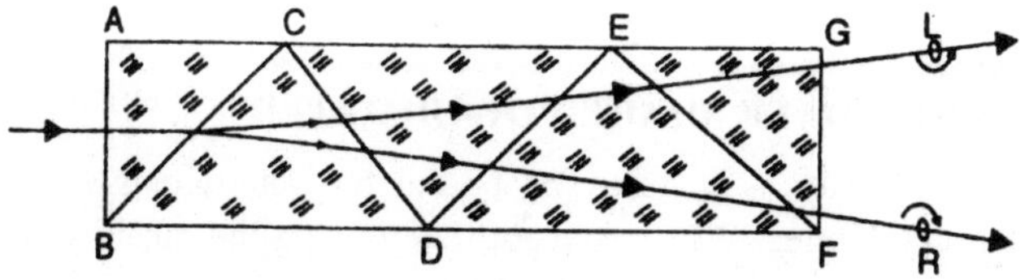

Fig. 8

When the two waves are analysed by a quarter wave plate and a Nicol prism, both are found to be circularly polarised in opposite directions. Thus, Fresnel's hypothesis is verified.

SPECIFIC ROTATION

If an optically active material is kept between two crossed polarizers, the field of view becomes bright. In order to get darkness once again, the analyser has to be rotated through an angle. The angle through which the analyser is rotated equals the angle through which the plane of polarization is rotated by the optically active substance. This angle depends on

(a) the thickness of the substance,

(b) density of the material or concentration of the solution,

(c) wavelength of light, and

(d) the temperature.

The amount of rotation θ caused by crystalline materials is given by

$$\theta = al$$

In solutions the amount of rotation θ is given by

$$\theta = scl$$

where c is the concentration and s is called the specific rotation.

The specific rotation for a given wavelength of light at a given temperature is defined conventionally as the rotation produced by one decimetre long column of the solution containing 1 gm of optically active material per c.c .of solution.

$$[S]_\lambda^t = \frac{\theta}{l \times C} = \frac{\text{Rotation in degrees}}{\text{Length in decimeters} \times \text{conc. in gm/c.c.}}$$

$$= \frac{10\theta}{l(\text{cm})C}$$

LAURENT'S HALF SHADE POLARIMETER

A polarimeter is an instrument used for determining the optical rotation of solutions. When used for determining the quantity of sugar in a solution it is called a *saccharimeter*.

Construction : A polarimeter consists of a glass tube for holding the solution under test held between crossed Nicol prisms, N_1 and N_2. Beyond the polarising Nicol prism a half-shade plate is located which is used for accurately adjusting the two Nicol prisms for crossed position. G is a glass tube which contains the optically active solution. Light from a monochromatic source is rendered parallel by the lens L and is incident on the polarizer, N_1. The light transmitted by the polarizer is plane polarised. The polarised beam then passes through the half-shade plate and then through the glass tube G. The light emerging from the solution will be incident on the analyser N,. The light is observed through a telescope T. The analysing Nicol N_2 can be rotated about the axis of the tube and the rotation can be measured with the help of a graduated circular scale.

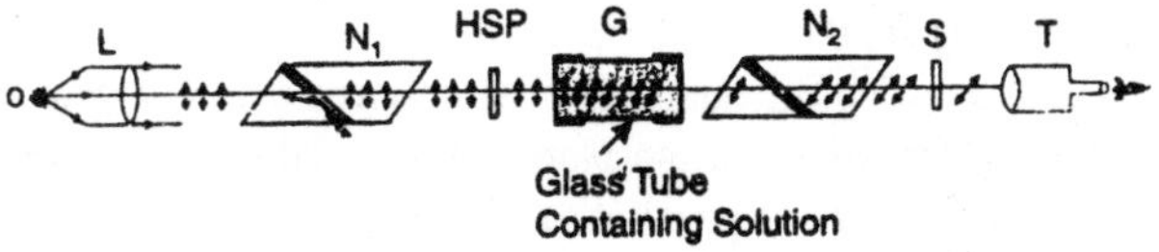

Fig. 9

Working : To find the specific rotation of a solution, the analyser is first adjusted such that field of view is completely dark. Then the glass tube is filled with the solution and is held in position. The field of view now becomes illuminated. The field of view can be again be made dark by rotating the analyser through a certain angle which gives the optical rotation of the solution.

The practical difficulty in this method is la, determination of the exact position for which complete darkness is achieved. The difficulty is overcome by using what is known as a *Laurent's half-shade device.* It consists of a semicircular half wave plate ACB of quartz cemented to a semicircular plate ADB of glass.

The optic axis of the wave plate is parallel to the line of separation AB. The half wave plate introduces a phase difference of 180° between e-ray and o-ray passing through it. The thickness of the glass plate is such that it transmits the same amount of light as done by the quartz half wave plate. One half of the incident light passes through the quartz plate ACB and the other half through the glass plate ADB. The light after passing through the polarizer is incident normally on the half shade plate and has vibrations along OR On passing through the glass, half the vibrations will remain along OP but on passing through the quartz half, the vibrations will split into e-and o-rays.

The o-vibrations are along OD and e-vibrations are along OA. The half wave plate introduces a phase difference forward between the two vibrations. The vibrations of o-ray will occur along OC instead of OD on emerging from the plate. Therefore the resultant vibration will be along OQ whereas the vibrations of the beam emerging from glass plate will be along OR In effect, the half wave plate turns the plane of polarisation of the incident light through an angle 2θ.

If the principal plane of the Nicol N_2 is aligned parallel to OP, the plane polarized light emerging from the glass tube will pass through the glass plate of the half shade plate and that part appears brighter. On the other hand light coming out of the quartz plate is partially obstructed and the corresponding field of view appears less bright. If the principal plane of N_2 is parallel to OQ the quartz half will appear brighter than the glass half. Thus, the two halves of the plate are unequally illuminated. When the principal plane of N_2 is parallel to AB, the two halves appear equally bright and when it is parallel to CD, the two halves are equally dark.

To find the specific rotation of a solution, the analyser is first set in the position for equal darkness without solution in the tube G. The reading on the circular scale is noted. Next, the tube is filled with the optically active solution of known concentration. The field of view is now partially illuminated. The analyser is rotated till the field of view becomes equally dark. The reading on the circular scale is noted again.

The difference between the two scale readings gives the angle of rotation of the plane of polarisation caused by the solution. Knowing the values of θ, l, and c, the specific rotation is obtained using the formula (44). Or otherwise, knowing the value of the specific rotation, the concentration of the solution can be determined with the help of the equation (44).

In the actual experiment, different concentrations of solutions are taken and the corresponding angles of rotation are determined. A graph is plotted between concentration C and the angle of rotation θ. The graph is a straight line (Fig. 10).

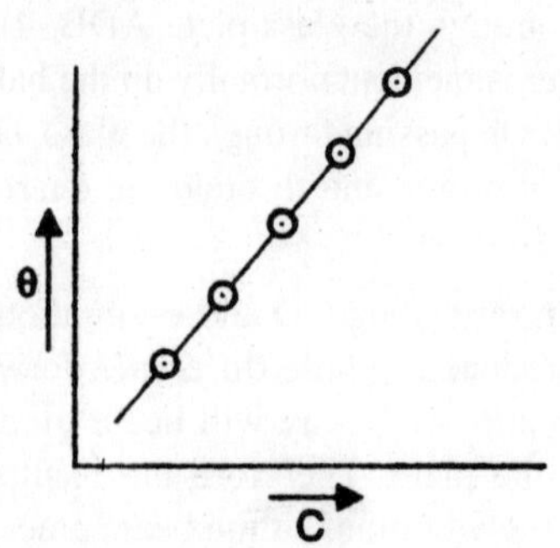

Fig. 10

Using the value of the slope in eqn. (44), the specific rotation of the optically active substance is calculated.

LIPPICH POLARIMETER

The schematic of the polarimeter is shown in Fig. 1.61. It consists of two Nicol prisms N_1 and N_2. Behind N_1, there is a Nicol prism N_3 that covers half the field of view. The Nicols N_1 and N_3 have their planes of vibration inclined at a small angle. Suppose the plane of vibration of N_1 is along AB and that of N_3 is along CD. The angle between the two planes is θ. When the analyser N_2 is rotated such that the plane of vibration of N_2 is along AB, the left half will be brighter than the right half. If the analyser N_2 has its plane of vibration along CD, the right half will be brighter than the left half. YY' is the bisector of the angle AOC.

Therefore, when the plane of vibration of the analyser N_2 is along YY', the field of view is equally illumined. For as light rotation of the analyser, either to the right or to the left, the field of view appears to be of unequal brightness. Therefore, by rotating N_2 the position for equal brightness of the field of view is obtained.

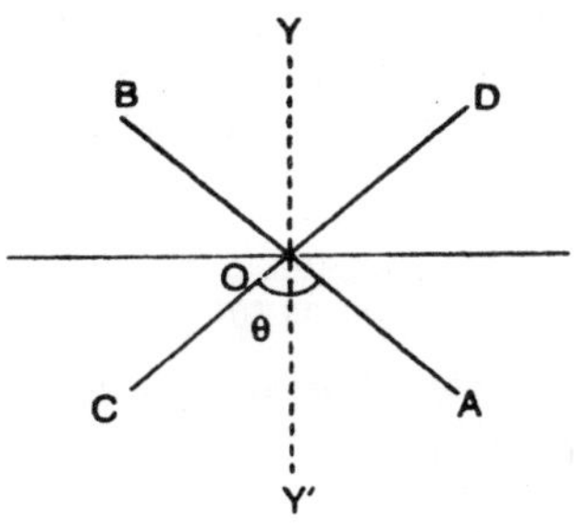

Fig. 11

Three-field system : In the improved form, Lippich polarimeter has a three-field system. The defect in the two field system is that if the eye is off the axis, even for the position of equal brightness of the field of view, one side appears brighter than the other. Just behind N_1 there are two Nicol prisms N_3 and N_4, Fig. 11.

The planes of vibration of N_3 and N_4 are parallel to each other and make a small angle with the plane of vibration of N_1. For a particular position of the analyser N_2 the field has three parts. The central portion is illuminated by light, which has passed through N_1 and N_2 while the other two portions, which are equally bright, are illuminated by light passing through N_1, N_2 and one of the prisms N_3 and N4.

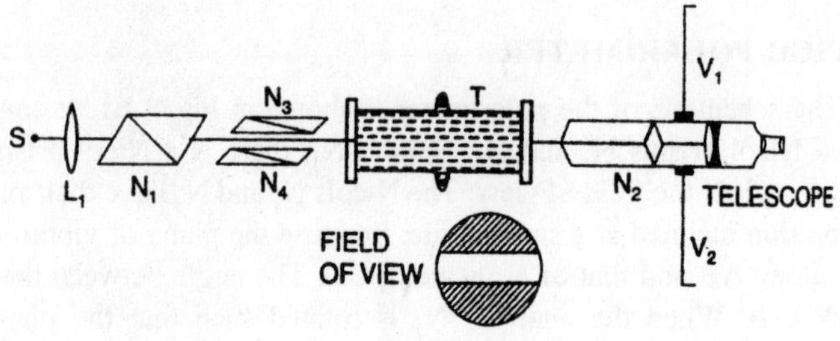

Fig. 12

ARTIFICIAL DOUBLE REFRACTION

Isotropic transparent materials such as glass do not exhibit double refraction under ordinary circumstances. However, they acquire the optical properties of a umaxyl crystal under the action of external forces. Consequently, they exhibit double refraction. The appearance of double refraction under the influence of an external agent is known as *artificial double refraction* or *induced birefringence.* The direction of the optical axis in such materials will be collinear with the direction of the external force. The action of the external force is to cause distortion of the molecular arrangement within the material and thereby transform the isotropic substance into an anisotropic substance. The induced birefringence disappears as soon as the external force ceases to act.

The property of double refraction can be induced in an isotropic material under the action of

(i) a mechanical strain

(ii) an electric field or

(iii) a magnetic field.

The materials which experience a change in their optical behaviour under the action of an electric field are called *electro-optic materials* and the resulting optical effects are known as *electro-optic effects.* Similarly, the materials that get influenced by a magnetic field are ca.led magneto-optic materials and the resulting optic effects are known as magneto-optic effects. The *electro-optic materials* play a very important role in modern technology.

Anisotropy Induced by Mechanical Strain

Materials such as glass become double refracting when they are subjected to mechanical strain. Experiments show that the induced birefringence is directly proportional to the stress σ experienced at a given point of the material. Thus,

$$\Delta\mu = k\sigma \qquad ...(1)$$

where k is the proportionality constant characteristic of the material. This effect was discovered by Sir David Brewster in 1816.

When the material is held between crossed polarisers P and A, as in Fig. 13, the field of view appears dark as long as the external force is not applied. As soon as the force is applied, coloured contours will be seen. Dark regions indicate the absence of strain in those areas. Each coloured contour shows the areas that are identically deformed. Such contours enable us assess the distribution of the stresses in the material. The mechanically induced bire-fringence is used to study stresses in girders, beams etc. The model of the object under investigation is made of transparent plastic material and is then loaded. Using crossed polarizer system, the stresses produced at different positions are analysed and estimated. This method of analysis is known as *photoelastic analysis* and is widely employed in civil and mechanical enginee-ring practices.

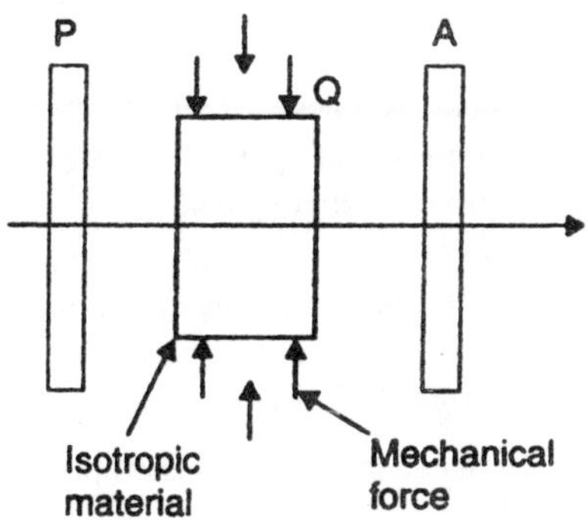

Fig. 13 : Birefringence is induced in an isotropic material under mechanical strain

Kerr Effect

Anisotropy induced in an isotropic liquid under the influence of an electric field is known as the Kerr effect. John Kerr discovered it in 1875.

A Kerr cell is required for studying the effect. It consists of a sealed glass cell filled with a liquid comprising of asymmetric molecules. Two plane electrodes of specific length *l* are arranged in it with their faces strictly parallel to each other.

When a voltage is applied to them a uniform electric field is produced in the cell. The Kerr cell is placed between a crossed polarizer system. When -the electric field is applied, the molecules of the liquid tend to align along the field direction. As the molecules are asymmetric, the alignment causes anisotropy and the liquid becomes double refracting. The induced birefringence is proportional to the square of the applied electric field and to the wavelength of incident light. Thus,

$$\Delta\mu = J\lambda E^2 \qquad ...(2)$$

where K is known as the Kerr constant.

Among the liquids, nitrobenzene ($C_6H_5NO_2$) is found to have the highest value for the Kerr constant. Therefore, Kerr cells use nitrobenzene.

POCKELS EFFECT

This is also an electro-optic effect. F. Pockels discovered in 1893 that the application of an electric field to piezoelectric crystals makes them birefringent. Normally, piezoelectric crystals are birefringent but in certain directions do not exhibit double refraction. When an electric field is impressed along these directions, double refraction is induced along these directions also. A Pockels cell consists of a piezoelectric crystal, for example lithium niobate placed between crossed polarisers. Transparent electrodes (thin conducting coatings of tin oxide or indium) are deposited on opposite sides of the crystal. The crystal is oriented in such a way that its optic axis lies along the direction of the electric field applied between the electrodes. The transparent electrodes ensure free propagation of light through the crystal.

The birefringence induced in the crystal is proportional to the strength of the applied field. Thus,

$$\Delta\mu = kE \qquad ...(3)$$

where k is a constant characteristic of the material. Eqn. (3) shows that Pockels effect is a linear effect.

The total birefringence of the cell is initially made equal to λ/1. When the electric field is increased, the beam is transmitted or hindered depending on the phase difference between the o-ray and e-ray. The device switches on and off periodically. Pockels cells are used in fast switching applications and in fibre optics. It can be used to obtain amplitude, frequency or phase modulation.

A Pockels cell is simple in construction and requires a small voltage of the order of 1.5 kV where as a Kerr cell is complicated in construction and requires higher voltages of the order of 15 kV. The piezoelectric crystals of ammonium dihydrophosphate (ADP) and potassium dihydrophosphate (KDP) are widely used in Pockels cell. Kerr and Pockels cells are widely used as electro-optic shutters for Q-switching of lasers.

COTTON-MOUTON EFFECT

The Cotton-Mouton effect is a magneto-optic effect. An isotropic material acquires the optical behaviour of a uniaxial crystal under the action of an external magnetic field. Cotton-Mouton effect. Birefringence is induced in a material subjected to a magnetic field.

The induced birefringence is governed by the relation

$$\Delta\mu = C\lambda B^2 \quad \text{...(1)}$$

where C is a constant characteristic of the material. The magnitude of the induced birefringence is usually very small.

Faraday Effect

Optically inactive substances acquire the ability of rotating the plane of polarisation of light when they are subjected to a magnetic field, parallel to propagation direction. Michael Faraday discovered this effect and hence it is called Faraday effect.

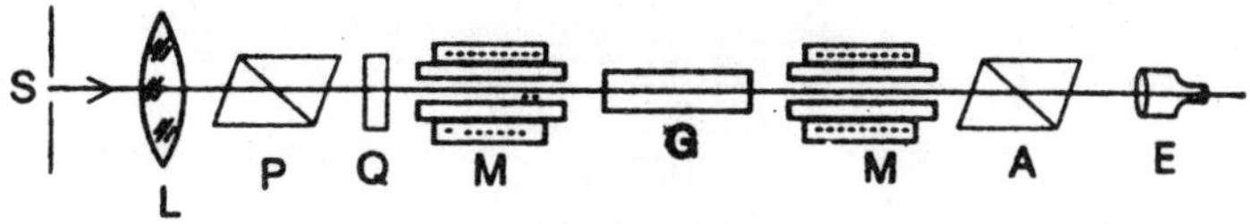

Faraday effect

Fig. 14

The angle of rotation θ of the plane of polarisation is proportional to the length of the path of light in the material and to the strength of the applied magnetic field. Thus,

$$\theta = VlH \qquad ...(2)$$

where V is known as Verdet constant.

The angles θ of rotation are not very large. For magnetic field strengths of the order of 10^6A/m and $l = 0.1$ m, θ is about 1° to 2°. One of the interesting problems encountered in satellite commu-nications is the Faraday effect. As radio waves pass through the ionosphere, their plane of polarisation is rotated by the ionised particles in conjunction with the Earth's magnetic field. A horizontally polarised wave becomes vertically polarised because of the Faraday rotation in the ionosphere. The reception problem is solved by using an antenna with circular polarisation, which ensures that the waves are received satisfactorily, no matter how they have been rotated.

LCDs

Liquid crystal display (LCD) devices constitute one of the most beautiful applications of polarisation phenomenon. LCDs are widely used as alphanumerical read-outs in wristwatches, calculators, clocks, electronic instruments, video games, etc. An LCD consists of a liquid crystal material, which is double refracting, of about 10 μm thick suitably supported between two thin glass plates having transparent conducting coatings on their inner surfaces (Fig. 14a). The conducting coating is etched in the form of a digit or character, as shown in (Fig. 14b). The assembly of glass plates with liquid crystal material in between is sandwiched between two crossed-polarizer sheets.

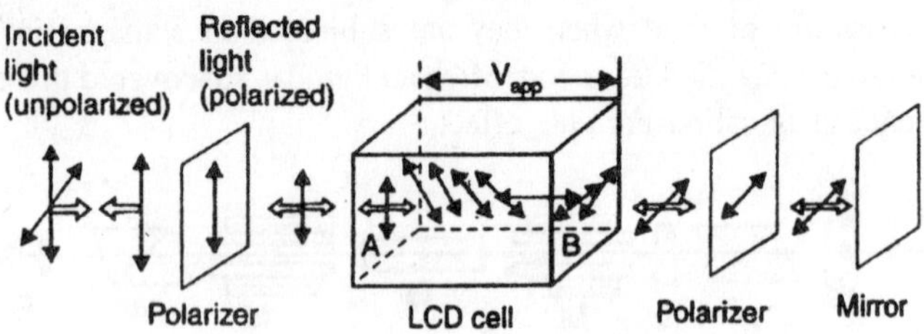

Fig. 15

During the fabrication of LCDs the liquid crystal molecules are aligned in such a way that their long axes undergo a 90° rotation, from plate A to plate B. It is called a twisted molecular arrangement. When natural light is incident on the assembly, the front polarizer converts it into linearly polarised light. As the linearly polarised light propagates through the LCD, the optical vector is rotated through 90° by the twisted molecular arrangement. Therefore, it passes unhindered through the rear polarizer whose transmission axis is perpendicular to that of the front polarizer. A reflecting coating at the back of the rear polarizer sends back the light, which emerges unobstructed by the front polarizer. Consequently, the display appears uniformly illuminated. When a voltage is applied to the device, the molecules between the elecodes untwist and align align the field direction. As a result, the optical vector does not undergo rotation as it passes though that regin.

POLARIZER AND ANALYSER

A polarizer is an optical device that transforms unpolarized light into polarized light (Fig. 16). If it produces linearly polarised light, it is called a linear polarizer. A linear polarizer is associated with a specific direction called the transmission axis of the polarizer. If natural light is incident on a linear polarizer, only that vibration that is parallel to the transmission axis is allowed through the polarizer while the vibration that is in a perpendicular direction is totally blocked. An analyser is a device, which is used to identify the direction of vibration of linearly polarised light. A polarizer and an analyser are fabricated in the same way and have the same effect on the incident light.

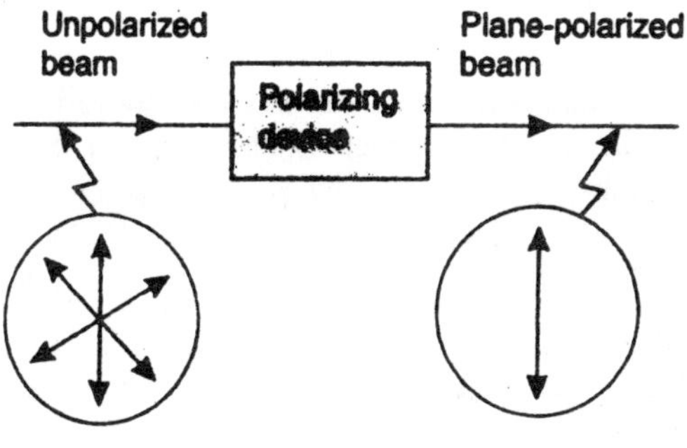

Fig. 16 : A polarizer transforms unpolarized light into polarized light.

Fabrication of Linear Polarizer

Practical linear polarizers are fabricated using birefringent crystals or dichroic crystals. Out of the polarisers based on double refraction, Nicol prism is the widely used one. Polarizer sheets made using dichroism are known as Polaroid sheets. We describe here the preparation of Nicol prism and Polaroid sheets.

(i) *Nicol Prism :* A Nicol prism is made from calcite crystal. It was designed by William Nicol in 1820. A rhomb of calcite crystal about three times as long as it is thick, is obtained by cleavage from the original crystal. The ends of the rhombohedron are ground until they make an angle of 68° instead of 71° with the longitudinal edges. This piece is then cut into two along a plane perpendicular both to the principal axis and to the new end surfaces. The two parts of the crystal are then cemented together with Canada balsam, whose refractive index lies between the refractive indices of calcite for the o-ray and e-ray. $\mu_o = 1.66$, $\mu_e = 1.486$ and $\mu_{canada\ baisam} = 1.55$. The position of the optic axis AB. The refractive index for e-ray depends upon the direction in which e-ray is propagating in the crystal. The difference between the refractive index of o-ray and that for e-ray goes on increasing with the angle between the two rays in the crystal. When this angle is 90°, the difference is a maximum. Thus, for a fixed value for μ_o, the μ_e has its maximum or minimum value in perpendicular direction. In the above case $\mu_e = 1.486$ represents the minimum value.

The ray after entering the crystal suffers double refraction and splits up into o-ray and e-ray. The two rays with their directions of vibrations. The values of the refractive indices and the angles of incidence at the Canada balsam layer are such that the e-ray is transmitted while the o-ray is internally reflected. The face where the o-ray is incident is blackened so that the o-ray is completely absorbed. Then we get only the linearly polarised e-ray coming out of the Nicol with direction of vibration as shown in the Fig. 2.68. Thus, the Nicol works as a polarizer.

The Nicol prism is the most widely used polarizing device. Nicols are good polarizers but they are expensive and have a limited field of view of about 28°.

For studying the optical properties of transparent substances, two Nicols are used–one as polarizer and the other as analyser.

(ii) *Polaroid Sheets :* In 1928 E.H. Land invented a method of aligning small crystals to obtain large polarising sheets. The sheets are called Polaroid sheets. They are fabricated as follows. A clear plastic sheet of long chain molecules of PVA (polyvinyl alcohol) is heated and then stretched in a given direction to many times its original length.

During the stretching process the PVA molecules become aligned along the direction of stretching. The sheet is then laminated to a rigid sheet of plastic to stabilise its size. It is then exposed to iodine vapour. The iodine atoms attach themselves to the straight long chain PVA molecules and consequently form long parallel conducting chains. The conduction electrons associated with iodine can move along the chains.

When natural light is incident on the sheet, the component that is parallel to the chains of iodine atoms induces current in the conducting chains and is therefore strongly absorbed. Consequently, the light transmitted contains only the component that is perpendicular to the direction of molecular chains. The direction of E-vector in the transmitted beam corresponds to the *transmission axis* of the Polaroid sheet. These sheet polarisers are inexpensive and can be made in large sizes. Polaroid sheets are widely used in sunglasses, camera filters etc. to eliminate the unwanted glare from objects. They are used respectively as polarisers and analysers to produce polarised light and to examine the state of polarisation of light.

Working

Nicol prisms and Polaroid sheets are widely used for the production and detection of linearly polarised light. These components are appropriately mounted in metal ring structures such that the transmission axis can be rotated and the amount of rotation may be read from the graduations marked on the ring.

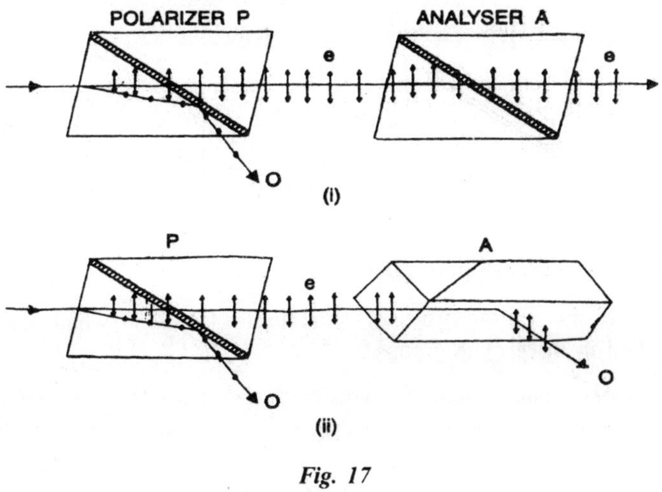

Fig. 17

When two Nicol prisms P and A are placed adjacent to each other as shown in Fig. 17, one of them acts as a polarizer and the other acts as an analyser. If an unpolarized ray of light is incident on the Nicol prism P, a linearly polarised e-ray emerges from P with its vibration direction lying in the principal section of P.

The state of the polarisation of the light emerging from polariser P can be examined with another polarizer A, which for convenience is called an analyser. Let now this ray be incident on the second Nicol prism A, whose principal section is parallel to that to P. The vibration direction of the ray will be in the principal section of A and hence it is transmitted unhindered through the analyser A.

If the Nicol prism A is gradually rotated, the intensity of the e-ray decreases in accordance. When its principal section becomes perpendicular to that of the Nicol prism P, the vibrations of the ray, emerging from P and incident on A, will be perpendicular to the principal section of A. In this position the ray behaves as o-ray inside the prism A and is totally internally reflected by the Canada balsam layer.

Hence no light is transmitted by the prism A. In this configuration, the two Nicol prisms P and A are said to be *crossed.* If the Nicol prism A is further rotated through another 90°, the intensity of light emerging from A will go on increasing. The intensity will become a maximum when its principal section is again parallel to that of the prism P. Thus, the prism P produces linearly polarised light while the prism A detects it. Hence the prism P is called a polarizer and .the prism A an analyser.

DEMONSTRATION OF POLARISATION

The polarisation effect can be very easily demonstrated by connecting an electric bell across two electrodes of a simple cell. When the cell works, the sound of the bell will be loud at first but it becomes fainter and fainter with time and finally it will stop when the cell is completely polarised.

ELIMINATION OF POLARISATION

There are various ways of removing polarisation which can be classified in three main groups :

(i) *Mechanical Method :* This method depends upon the removal of the hydrogen in a mechanical manner, *i.e.*, brushing off. If this be done every now and then, no layer of hydrogen will permanently deposit upon the plate, and the cell will not become polarised. But this is a laborious task. *Smee* suggested that if the surface be made rough, the points on the surface will not allow hydrogen bubble to accumulate upon the plate.

(ii) *Electrochemical Method :* This is one of the most useful method for reducing the polarisation. In this method two solutions are employed so that the liberated hydrogen comes in contact with the second solution from which ions of the metal as that of the positive electrode are liberated (cf. Daniell cell) or some such gas is liberated which does not cause polarisation (cf. Bunsen's cell).

(iii) *Chemical Method :* The polarisation of the cells may be minimised by using some strong oxidising agents such as chromic acid, nitric acid, manganese dioxide, etc. which may convert hydrogen into water, according to the reaction.

$$H_2 + 1/2O_2 \rightarrow H_2O$$

The substance which removes the polarisation completely or partly is known as depolariser. Some examples of depolarisers are :

(i) Solid MnO_2 acts as depolariser in the Leclanche cell.

(ii) Oxygen acts as a depolariser in the rusting of ion, which is facilitated by the presence of air.

Effect of Polarizer on Natural Light

When unpolarized light passes through a polarizer, the intensity of the transmitted light will be exactly half that of the incident light. We can prove this as follows.

Let E_0 be the amplitude of vibration of one of the waves of the unpolarized light incident on the polarizer. Let E_0 make an angle θ with the transmission axis of the polarizer. E_0 may be resolved into its rectangular components E_x and E_y as shown in Fig. 18.

The polarizer blocks the component E_x and transmits the component E_y. The intensity of the transmitted light is then

$$I = E^2_y \ (\cos \theta)^2 = E^2_0 \cos^2 \theta = I_0 \cos^2 \theta \qquad ...(1)$$

In unpolarized light, all values of Q starting from 0 to In are equally probable.

$$\therefore \quad I = I_0 \{\cos^2 \theta\} \qquad ...(2)$$

$$= \frac{I_0}{2\pi} \int_0^{2\pi} \cos^2 \theta \, d\theta = \frac{I_0}{2\pi} \int_0^{2\pi} \left(\frac{1 + \text{cis } 2\theta}{2} \right) d\theta$$

$$= \frac{I_0}{4\pi} \left[(\theta)_0^{2\pi} + \left\{ \frac{\sin 2\theta}{2} \right\}_0^{2\pi} \right]$$

$$= \frac{I_0}{4\pi} (2\pi = 0)$$

$$\therefore \quad I = \frac{1}{2} I_0$$

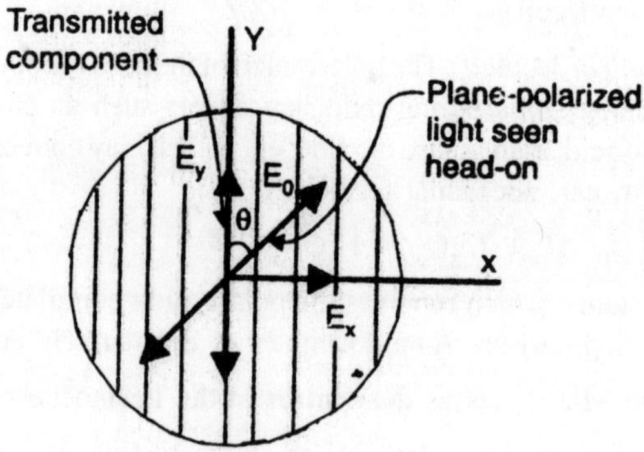

Fig. 18 : Action of a polarizing sheet on a linearly polarized wave oriented at an angle θ w.r.t. the transmission axis. The component E_X is absorbed and E_Y component is transmitted.

Thus, if unpolarized light of intensity I_0 is incident on a polarizer, the intensity of light transmitted through the polarizer is $\frac{I_0}{2}$.

Effect of Analyser on Plane Polarized Light—Malus' Law

When natural (unpolarized) light is incident on a polarizer, the transmitted light is linearly polarized. If this light further passes through an analyser, the intensity varies with the angle between the transmission axes of the polarizer and analyser. Malus studied the phenomenon in 1809 and formulated the law that bears his name.

It states that the *intensity of the polarized light transmitted through the analyser is proportional to cosine square of the angle between the plane of transmission of the analyser and the plane of transmission of the polarizer. This is known as Malus' law.*

If unpolarized light of intensity I_o is incident on a polarizer, plane polarized light of intensity $\frac{I_o}{2}$ is transmitted by it. Let us denote $\frac{I_o}{2}$ by I_1. This plane polarized light then passes through the analyser. Let E be the amplitude of vibration and θ be the angle that this vibration makes with the axis of the analyser. E can be resolved into two rectangular components: (i) E_y, parallel to the plane of transmission of the analyser and (ii) E_x, perpendicular to the plane of analyser. It is only the parallel component E that is transmitted by the analyser.

$$E_y = E \cos \theta$$

The intensity corresponding to this component is

$$I = E^2 \cos^2 \theta = I_1 \cos^2 \theta = \left(\frac{I_o}{2}\right) \cos^2 \theta$$

Case (i): If $\theta = 0°$	axes parallel	$I = I_1$
Case (ii): If $\theta = 90°$	axes perpendicular	$I = 0$
Case (i): If $\theta = 180°$	axes parallel	$I = I_1$
Case (i): If $\theta = 270°$	axes perpendicular	$I = 0$

Thus, we obtain two positions of maximum intensity and two positions of zero intensity when we rotate the axis of the analyser with respect to that of the polarizer.

Concentration Polarisation

As the current passes through the cell, there occurs concentration changes at the electrodes and these changes develop a back E.M.F. This changes the resultant E.M.F. of the cell. *Thus, the phenomenon of the departure of the electrode potential (increase or decrease) from the reversible value as a result of the change of concentration in the vicinity of the electrodes is known as concentration polarisation.*

The concentration polarisation arises due to the slow movement of the ions towards or away from the electrode. This effect can be minimised by stirring the electrolyte, raising the temperature and by increasing the concentration. This is, however, not completely eliminated because diffusion of ions is a slow process.

SOLVED EXAMPLES

Example 1:

The E.M.F of the following cell is 0.086 volt at 25°C.

$$Ag \left| \begin{matrix} 0.0093\ N \\ AgNo_3 \end{matrix} \right\| \begin{matrix} NH_4NO_3 \\ sat. \end{matrix} \left\| AgNO_3(x) \right| Ag$$

Find the concentration (x) of the unknown solution.

Solution:

$$E = \frac{2.303\ RT}{NF} \log \frac{C_2}{C_1} = 0.00591 \log \frac{x}{0.0093}$$

$$0.086 = 0.0591 \log \frac{x}{0.0093}$$

$$\text{or} \quad \log \frac{x}{0.0093} = \frac{0.086}{0.0591} = 1.390.$$

$$\therefore \quad \frac{x}{0.0093} = \text{Antilog } 1.390 = 24.45$$

$$\text{Hence} \quad x = 0.0093 \times 24.45 = 0.228.$$

Example 2:

The E.M.F. of the cell

10% Zn in Hg | $ZnSO_4$. $7H_2O$: Hg_2SO_4 (s) | Hg

is given by the reaction $E_t = 1.4328 – 0.00119\ (t–15) – 0.000007\ (t–15)^2$ volts. Formulate the reaction and find the values of ΔG and ΔH for the above cell at 25°.

Solution:

The cell's reaction is given by

$$7H_2O + Zn + Hg_2SO_4\ (s) \rightarrow 2Hg + ZnSO_4 .\ 7H_2O$$

At 25°C, the E.M.F. of the cell is

$$E25^\circ = 1.4328 - 0.00119\ (25 - 15) - 0.000007\ (25 - 15)^2 \text{ volts}$$

$$= - 1.4262 \text{ volts.}$$

$$\text{From equation } \Delta G = - nEF^* = - \frac{2 \times 1.4202 \times 96500}{4.185}$$

$$\Delta G = - 65500 \text{ calls.}$$

From Gibb's Helmholtz equation ΔG – ΔH

$$= -\,nTF^{*}\,(\partial E/\partial T)_p \qquad ...(1)$$

For finding out the value of $(\partial E/\partial T)_{p,}$ differentiate the relation between E and t° with respect to temperature.

Then

$$(\partial E/\partial T)_p = 0 - 0.00119 - 2 \times 0.000007\,(t - 15)$$

At $\quad 5°C\ (\partial E/\partial T)_p = [\,-0.00119 - 0.000014]\,(25 - 15)$

$$= -\,0.00119 - 0.000014 = -\,0.00133 \text{ volt/degree.}$$

Therefore, from (1), we get

$$-\,65500 - \Delta H = -\,\frac{298.16 \times 96500 \times 2\,(-\,0.00133)}{4.185}$$

$\Delta H = -\,8.3790$ calls.

Example 3:

At a particular incidence angle the coefficient of amplitude reflection of glass plate is 2% for | | vibrations and 8% for ⊥ vibrations. Calculate the degree of polarization of beam passing through a pile of 10 plates at this incidence angle, neglecting absorption.

Solution:

10 plates have 20 surfaces. At each surface 98% of one and 92% of the other vibration is transmitted (in amplitude). If a is the initial amplitude in each plane (| | and ⊥ to plane of incidence), the emergent beams have amplitudes

$$a_1 = ax\,(0.98)^{20} = .667a,$$

$$a_2 = ax\,(0.92)^{20} = .180a.$$

The corresponding intensities are (on squaring)

$$I_1 = I_0 x\ 0.445$$

$$I_2 = I_0 x\ 0.036$$

$$\text{Degree of Polarization} = \frac{I_{max} - I_{min}}{I_{max} + I_{min}} = \frac{0.445 - 0.036}{0.445 + 0.036} = 0.85.$$

Example 4:

The E.M.F. of the cell at 25°C

$$H_2 \mid 0.5N - HCl \mid 0.1\ N - NaOH \mid H_2$$

is 0.725 volt. H_2 at each electrode is under atmospheric pressure and diffusion potential is neglected. What is the ionic product of water if the dissociation of 0.5N HCl = 0.87 and of 8.1 N NaOH = 0.9 ?

Solution:

The E.M.F. of the above cell is given by

$$E = \frac{RT}{nF} \log \frac{\text{Concentration of } H^+ \text{ in HCl}}{\text{Concentration of } H^+ \text{ in NaOH}}$$

$$= 0.0591 \log \frac{C_{H^+} \text{ in HCl}}{C_{H^+} \text{ in NaOH}}$$

$C_H{}^+$ in 0.5 N – HCl = 0.5 × 0.87 = 0.435 gm/ion.

$$C_H{}^+ \text{ in } 0.1 \text{ N NaOH} = \frac{C_{H^+} \times C_{OH^-}}{C_{OH^-}} = \frac{k_w}{0.1 \times 0.9} = \frac{k_m}{0.09}$$

$$\therefore \quad 0.725 \text{ volt.} = 0.059 \log \frac{0.435 \times 0.09}{k_w}$$

or

$$\log \frac{0.03915}{k_w} = \frac{0.725}{0.0591} = 12.27$$

$$k_w = 0.03915/\text{Antilog } 12.27 = 0.03915/1.862 \times 10^{12}$$

$$= 2.102 \times 10^{-14}$$

Example 5:

The E.M.F. of the cell

Cd | $CdCl_2.2.5H_2O$ (satd) : AgCl(s) – Ag

is 0.6753 volt at 25°C and 0.6915 volt at 0°C. Calculate the heat of reaction at 25°C and find out the cell reaction.

Solution:

The cell reaction occurring is given by

$$Cd + 2AgCl(s) + aq \rightarrow CdCl_2\ 2.5\ (H_2O)\ (satd) + 2Ag$$

Evidently, this reaction takes place for the passage of two Faradays of electricity

We know that $$\Delta H = nF^*E - nF^*T \left(\frac{\partial E}{\partial T}\right) \quad ...(1)$$

Since the E.M.F. decreases with increase in temperature, therefore, $\left(\frac{\partial E}{\partial T}\right)$ is negative.

$$\left(\frac{\partial E}{\partial T}\right)_p = -\frac{(0.6915 - 0.6753)}{25} = -0.00065 \text{ volt / deg.} \qquad ...(2)$$

From (1), we have

$$-\Delta H = 2 \times 96500 \times 0.6753 - 2 \times 96500\ \ 298\ (-0.00065)$$
$$= 167717 \text{ joules}$$
$$= \frac{167717}{4.185} \text{cals} = 40050 \text{ cals}$$

Example 6:

An analyzer examines two adjacent plane polarized beams A, B whose planes of polarization ae mutually perpendicular. In one position of the analyzer, beam B shows zero intensity. From this position a rotation of 27° shows the two beams as matched. Deduce the intensity ratio IA/IB of the beams.

Solution:

In the zero position, as described, the vibrations transmitted by the analyzer are parallel to the vibrations in beam. A Hence, in the position 27° from this,

$$I_A \cos^2 27° = I_B \cos^2 (90° - 27°) = I_B \sin^2 27°$$
$$I_A/I_B = \tan^2 27° = 0.26$$

(This principle is used in the intensity comparison in many instruments).

Example 7:

Calculate the angular deviations of E and O rays passing through a 60° prism of $NaNO_3$ in minimum deviation condition, optic axis being perpendicular to the base.

Solution:

Using the equation above with Å = 60°

$$\sin \frac{1}{2}(60 + \delta_0) = 1.587 \times 0.5 = 0.794$$

$$\Rightarrow \qquad \delta_0 = 45°12'$$

$$\sin \frac{1}{2}(60+\delta_E) = 1.336 \times 0.5 = 0.668$$

$$\Rightarrow \quad \delta_E = 23°48'$$

[Note that the difference between μ_o and μ_E is much larger than the variations of μ with wavelength in most cases. For extra-dense flint glass difference of μ from violet to red light is less than 0.03].

Example 8:

Let a double image prism like be made of a material for which $\mu_0 = 1.662$, $\mu_E = 1.474$. Deduce the angular separation of the emergent beams.

Solution:

The angle of incidence at R is 45°. The angles of refraction are:

$$\sin^{-1}\left(\frac{1}{\sqrt{2}} \text{x} \frac{1.662}{1.474}\right) = 52°50';$$

$$\sin^{-1}\left(\frac{1}{\sqrt{2}} \text{x} \frac{1.474}{1.662}\right) \quad 38°\ 50'$$

These rays fall on the opposite face at angles of incidence + 7°50' and -6°10' respectively. The angles of emergence are

$$\sin^{-1} (1.474 \sin 7°50') = 11°34'$$

$$\sin^{-1} (1.662 \sin -6°10') = -\ 10°12';$$

The angle between the emergent beams therefore 21°46'.

EXERCISES

1. Determine the path and wave fronts of ordinary and extra ordinary rays inside a double refracting crystal on the basis of Huygens' theory when the optic axis is inclined to the incident surface.
2. What is a quarter wave plate? Deduce its thickness for a given λ in terms of its refractive indices.
3. Explain what is circular polarized light? How is it produced in laboratory with the help of a quarter wave plate?
4. Distinguish between polarised and unpolarized light. Discuss the process of production and detection of elliptically polarised light.

5. How can we experimentally distinguish between plane polarised, circularly polarised and elliptically polarised light?
6. What is Babinet's compensator? Explain how it can be used to analyse elliptically polarised light?
7. What is a Babinet's compensator? What are its advantages over a quarter wave plate?
8. Give the construction and working of Laurent's half shade polarimeter.
9. Give the construction and theory of Lippich polarimeter. How will you determine the specific rotation of a given solution with its help?
10. Explain Fresnel's theory of rotation of the plane of polarisation. How would you increase the sensitiveness of a pair of crossed Nicol prisms?
11. Show that for incidence at Brewster's angle, the angle between thel dent and refracted rays is 90°.

4
Polarization by Refraction

E.L. Malus, the French engineer and scientist discovered in 1808 polarisation of natural light by reflection from the surface of glass. He noticed that when natural light is incident on a plane sheet of glass at a certain angle, the reflected beam is plane polarized.

The perpendicular component is represented by dots and is called the *s-component*. The parallel component is represented by the arrows and is called the *p-component.* Fig. 1 shows an unpolarized light beam AB incident on a glass surface. The incident ray AB and the normal NBN' define the *plane of incidence*. The electric vector E of the ray AB can be resolved into two components, one perpendicular to the plane of incidence and the other lying in the plane of incidence. In case of completely unpolarized light the two components are of equal magnitude. When natural light is incident on the surface, the two components are reflected to different extents.

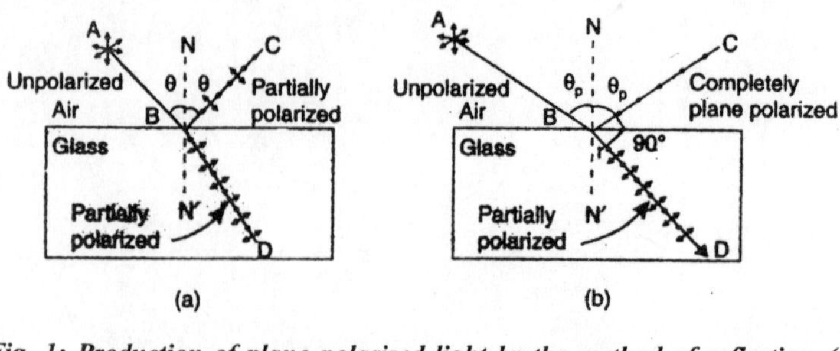

Fig. 1: Production of plane polarized light by the method of reflection–(a) When an unpolarized light is incident on a reflecting surface, the reflected and refracted beams are partially polarized, (b) The reflected beam is completely polarized when the angle of incidence equals the polarizing angle θ_p which satisfies Brester's law.

The reflected ray BC has a predominance of s-component (Fig. 1) and as such it is partially polarised. As the angle of the incidence of light is varied, the extent of polarisation of the reflected ray varies. At a particular angle θ_P the reflection coefficient for p-component goes zero and the reflected beam does not contain any p-component. It contains only s-component and is totally *linearly polarised.* The angle θ_P is called the *polarizing angle*. This particular method of obtaining linearly polarised light is not advantageous. The intensity of the reflected beam is very weak as only 15% of the s-component is reflected. In contrast, the refracted beam is strong and consists of a mixture of p-component, which is transmitted 100% and the balance 85% of s-component.

When unpolarized light is incident at Brewster angle on a smooth glass surface, the reflected light is totally polarized while the refracted light is partially polarized. If natural light is transmitted through a single plate, the transmitted beam is only partially polarized. If a stack of glass plates is used instead of a single plate, reflections from successive surfaces occur leading to the filtering of the s-component in the transmitted ray. Ultimately, the transmitted ray consists of p-component alone. If I_P and I_S are the intensities of the parallel and perpendicular components in the refracted light, the degree of polarisation of the transmitted light is given by where m is the number of plates required and 4 is the refractive index of the material.

$$P = \frac{I_P - I_S}{I_P + I_S} = \frac{m}{m + \left(\frac{2\mu}{1-\mu^2}\right)^2} \quad ...(1)$$

It is found that a stack of about 15 glass plates is required for this purpose. The glass plates are supported in a tube of suitable size and are held inclined at an angle of about 33° to the axis of the tube. Such an arrangement is called a *pile of plates*. Unpolarized light enters the tube and is incident on the plates at Brewster angle and the transmitted light will be polarised parallel to the plane of incidence.

However, this method of polarising light is also not efficient because a good portion of light is lost in reflections.

EFFECT OF POLARIZER ON TRANSMISSION OF POLARISED LIGHT

(i) If unpolarized light is incident on a polarizer, the transmitted light will be linearly polarised light. The intensity of the transmitted polarised light will be half the intensity of unpolarized light incident on the polarizer. The intensity of the transmitted light does not change on rotation of the polarizer.

(ii) If partially polarised light is incident on a polarizer, the intensity of the transmitted light will be dependent on the direction of the transmission axis of the polarizer. The intensity of the transmitted light will vary from a maximum value I_{max} to a minimum value I_{min} in one full rotation of the polarizer. Two positions of I_{max} and two positions of I_{min} occur in one complete rotation,

(iii) If plane polarised light is incident on the polarizer, the intensity of transmitted light varies from zero to a maximum value. Two positions of zero intensity and two positions of full intensity I occur in one complete rotation of the polarizer.

(iv) When circularly polarised light is incident on a polarizer, the intensity of the transmitted light stays constant in any position of the polarizer. The circular vibrations may be resolved into two mutually perpendicular linear vibrations of equal amplitude. When the circularly polarised light is incident on the polarizer, the vibrations parallel to its transmission axis pass through the polarizer while the perpendicular component is obstructed. When the polarizer is rotated, there is always a component of constant intensity parallel to the axis of the polarizer, which is freely transmitted. Hence, the intensity of the transmitted light is the same for all positions of the polarizer.

(v) In case of elliptically polarised light, the intensity of the light transmitted through the polarizer varies with the rotation of the polarizer from I_{max} to I_{min}. I_{max} is found when the polarizer axis coincides with the semi-major axis of the ellipse and I_{min} occurs when the polarizer axis coincides with the semi-minor axis of the ellipse.

RETARDERS OR WAVE PLATES

Retarders are a class of optical elements that serve to change the state of polarization of an incident wave. The operation of a retarder is very simple. When plane polarized light is incident on a retarder, it splits the light into two plane polarized light waves and one of the waves lags behind the other by a known amount. Upon emerging from the retarder, the two waves superpose on each other to produce a wave, which is of a different state of polarization. A quarter wave plate and a half wave plate are two important retarders. As calcite is brittle and difficult to handle in the form of thin slices, it is not generally used to make retardation plates. Retarders are frequently made from quartz but more often they are made using the biaxial crystal mica.

Quarter Wave Plate

A *quarter wave* plate is a thin plate of birefringent crystal having the optic axis parallel to its refracting faces and its thickness adjusted such that it introduces a quarter-wave ($\lambda/4$) path difference (or a phase difference of 90°) between the e-ray and o-ray propagating through it. When a plane polarized light wave is incident on a birefringent crystal having the optic axis parallel to its refracting face, the wave splits into e-wave and o-wave. The two waves travel along the same direction but with different velocities. As a result, when they emerge from the rear face of the crystal, an optical path difference would be developed between them. Thus, for a quartz wave plate,

$$(\mu_e - \mu_o)\, d = \frac{\lambda}{4} \qquad ...(1)$$

$$d = \frac{\lambda}{4[\mu_e - \mu_o]} \qquad ...(2)$$

A quarter wave plate introduces between e-ray and o-ray a phase difference δ given by

$$d = \frac{2\pi}{\lambda}\Delta = \frac{\pi}{2} = 90°$$

A quarter-wave plate is used in producing elliptically or circularly polarised light. It converts plane polarised light into elliptically or circularly polarised light depending upon the angle that the incident light vector makes with the optic axis of the quarter wave plate.

Action of quarter wave plate on elliptically and circularly polarised light:

Let us consider elliptically polarised light being incident on a quarter wave plate. Elliptically polarised light may be viewed as made up of two coherent plane polarised waves of different amplitudes, and differing in phase by 90°. the quarter wave plate introduces an additional phase difference of 90° leading to a total phase difference of 180° between the two component waves.

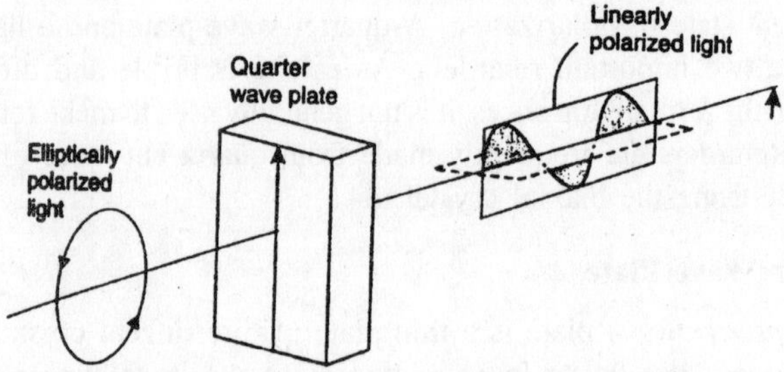

Fig. 2 : Action of a quarter wave plate. Linearly polarized light is produced in either case of elliptically or circularly polarized light incident on the plate.

When they emerge out of the plate, they combine to form linearly polarised wave. The action of the quarter wave plate on circularly polarised light wave is similar. Circularly polarised light incident on a quarter wave plate is converted into linearly polarised light.

Half Wave Plate

A half wave plate is a thin plate of birefringent crystal having the optic axis parallel to its refracting faces and its thickness chosen such that it introduces a half-wave ($\lambda/2$) path difference (or a phase difference of 180°) between e-ray and o-ray. When a plane polarized light wave is incident on a quartz crystal having the optic axis parallel to its refracting faces, it splits into two waves: o-and e-waves. The two waves travel along the same direction inside the crystal but with different velocities. As a result, when they emerge from the rear face of the crystal, an optical path difference would be developed between them.

$$(\mu_e - \mu_o)\, d = \frac{\lambda}{2} \qquad ...(3)$$

$$d = \frac{\lambda}{2(\mu_e - \mu_o)} \qquad ...(4)$$

A half wave plate introduces between e-ray and o-ray a phase difference δ given by

$$\delta = \left(\frac{2\pi}{\lambda}\right)\Delta = \pi = 180°$$

Rotation of the plane of polarisation of linearly polarised light by a half wave plate

Now let a plane polarized light be incident normally on the half-wave plate. Let the electric vector E make an angle θ with the optic axis of the half wave plate.

The incident wave splits into two waves, e-and o-waves. The waves progressively develop path difference as they travel through the crystal and they emerge with a phase difference of 180°. When the two waves combine, they yield a plane-polarised wave, which has its plane of polarization rotated through an angle of 2θ. Therefore, *a half-wave plate rotates the plane of polarization of the incident plane polarised light through an angle* 2θ. The half wave plate will invert the handedness of elliptical or circular polarised light, changing right to left and *vice versa.*

POLARIZATION BY SCATTERING

If a narrow beam of natural light is incident on a transparent medium containing a suspension of ultramicroscopic particles, the light scattered is partially polarized. The degree of polarisation depends on the angle of scattering. The beam scattered at 90° with respect to the incident direction is linearly polarized. The direction of vibration of E vector in the scattered light will be perpendicular to the plane defined by the direction of propagation and the direction of observation, as illustrated in Fig. 3. Sunlight scattered by air molecules is polarised. The maximum effect is observed on a clear day when the Sun is near the horizon. The light reaching an observer on the ground from directly overhead is polarised to the extent of 70% to 80%.

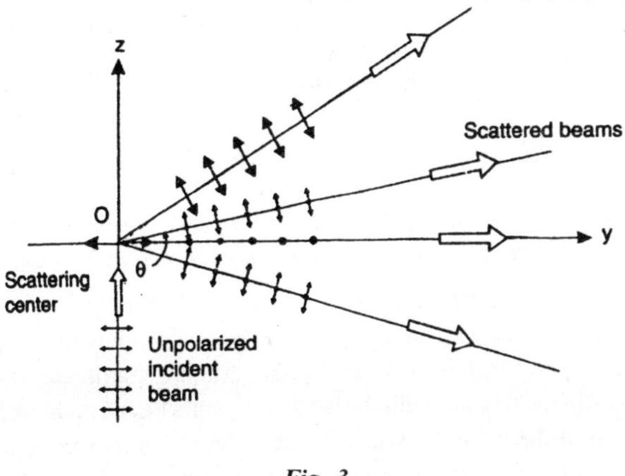

Fig. 3

HUYGENS' EXPLANATION OF DOUBLE REFRACTION

The propagation of light may be treated in terms of wave surfaces. If a point source of natural light is imagined to be embedded within an isotropic substance such as glass, it gives rise to one wave surface, which is spherical in shape. The wavefront stimulates atoms, which then act as sources of spherical wavelets, all of which are in phase. They expand in all directions with the same velocity. In case of a double refracting crystal, two wave surfaces will be formed simultaneously, with the result that the beam will split into two rays. The wave surface corresponding to o-ray propagates with the same velocity in all directions and is therefore spherical. The wave surface corresponding to e-ray is an ellipsoid of revolution about the optic axis, since e-ray travels with different velocities in different directions within the crystal.

The two wave surfaces touch each other at two points and the line joining these points defines the direction of the optic axis. As light propagates through the crystal, the two wave surfaces travel in different directions in the crystal. Ultimately, two refracted rays emerge from the crystal.

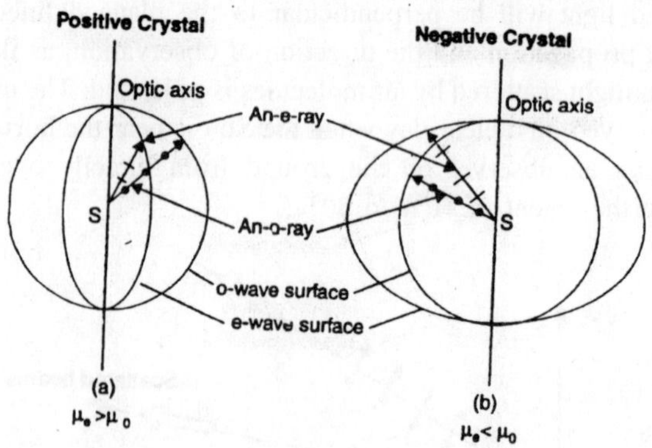

Fig. 4 : Huygens wave surfaces produced by a point source S-embedded in the birefringent crystal; (a) a positive crystal (b) a negative crystal.

Because of two different wave fronts, two different types of situations arise. The spherical wave front of o-ray may be enclosed by the ellipsoidal wave front of e-ray in one type of crystals. Such crystals are known as *negative crystals*. They are called negative crystals because the refractive index corresponding to the e-ray is less than that corresponding to o-ray. Calcite crystal is an example of negative type crystals. In the other case, the extraordinary wave front lies within the ordinary wave front and such crystals are called positive crystals. They are positive because the refractive index for the extraordinary ray is greater than that of o-ray.

o-Ray and e-Ray

We now compare the properties of o- and e-rays.

(i) Ordinary ray obeys the conventional laws of refraction, whereas the e-ray does not conform to them.

(ii) Both o-ray and e-ray are plane polarized. They are polarized in mutually perpendicular planes. The electric vector of o-ray vibrates perpendicular to the principal section of o-ray while the vibrations of e-ray take place parallel to the principal section of e-ray.

(iii) O-ray travels with the same speed in all directions within the crystal. The e-ray travels with different speeds along different directions in the crystal. However, the speed of e-ray will be equal to that of o-ray along the optic axis direction.

(iv) Because o-ray travels with the same velocity in all directions, the refractive index corresponding to it has a constant value. On the other hand, the refractive index for e-ray varies from direction to direction. The principal refractive index for o-ray is defined as follows:

$$\mu_o = \frac{c}{\upsilon_o} = \frac{\text{velocity of light in a vacumn}}{\text{Velocity of r - ray in the crystal}} \quad ...(1)$$

The principal refractive index for e-ray in positive crystals is defined as follows:

$$\mu_e = \frac{c}{(\upsilon_e)_{min}} = \frac{\text{velocity of light in a vacumn}}{\text{Minimum velocity of e - ray in the crystal}} \quad ...(2)$$

The principal refractive index for e-ray in negative crystals is defined as follows:

$$\mu_e = \frac{c}{(\upsilon_e)_{max}} = \frac{\text{velocity of light in a vacumn}}{\text{Maximum velocity of e - ray in the crystal}} \quad ...(3)$$

(v) When natural light is incident on an anisotropic crystal at an angle to the optic axis, it splits into o-and e-rays, which travel in different directions with different velocities.

When natural light is incident in a direction normal to the optic axis, o-ray and e-ray propagate in the *same direction* in the crystal but with *different velocities*. In a negative crystal e-ray leads o-ray and in case of a positive crystal o-ray leads e-ray.

NEGATIVE AND POSITIVE CRYSTALS

We compare here the characteristics of the positive and negative crystals.

(i) In positive uniaxial crystals, the ellipsoid of revolution cornspending to the e-ray is totally contained within the sphere corresponding to the o-ray. In negative uniaxial crystals, the ellipsoid of revolution for e-ray lies completely outside the sphere corresponding to o-ray. The two cases are depicted.

(ii) In positive crystals the e-ray velocity has a maximum value along the optic axis and a minimum value in a direction perpendicular to the optic axis. On the other hand, in negative crystals the velocity of e-ray has a minimum value parallel to the optic axis and a maximum value in a direction perpendicular to the optic axis.

(iii) In positive crystals, e-ray travels slower than o-ray in all directions except along the optic axis.

$\upsilon_e = \upsilon_o$ – parallel to optic axis

$\upsilon_e < \upsilon_o$ – other directions

In negative crystals, o-ray travels slower than e-ray in all directions except along the optic axis.

$\upsilon_e = \upsilon_o$ – parallel to optic axis

$\upsilon_e > \upsilon_o$ – other directions

(v) In positive crystals the principal refractive index for e-ray is larger than the principal refractive index for o-ray.

$$\mu_e > \mu_o$$

In negative crystals the principal refractive index for o-ray is larger than the principal refractive index for e-ray.

$$\mu_e < \mu_o$$

(v) *Birefringence* or amount of double refraction of a crystal is defined as

$$\Delta\mu = \mu_e - \mu_o \qquad ...(4)$$

$\Delta\mu$, is a positive quantity for positive crystals as $\mu_e > \mu_o$ in these crystals.

$\Delta\mu$ is a negative quantity for negative crystals as $\mu_e < \mu_o$ in these crystals.

CONSTRUCTION OF WAVEFRONTS BY HUYGENS'

A thinner rectangular cross-section of a crystal can be cut from a bigger crystal in three different ways. It can be cut in such a way that the optic axis lies :

(i) inclined to the refracting face

(ii) parallel to the refracting face

(iii) perpendicular to the refracting face. The path of o-ray and e-ray within a uniaxial crystal can be determined using Huygens' principle of secondary wavelets. We take the example of a negative crystal for the purpose of tracing the paths of o-ray and e-ray.

Case. 1. Optic axis inclined to the refracting edge

Let MNN' M' represent the calcite (negative) crystal. Let CD represent the monochromatic plane wave front incident normal to the crystal surface. The optic axis is *inclined to the refracting edge*. According to Huygens the phenomenon of double refraction involves two types of propagation of light waves. The wave known as o-wave has a spherical wave front whereas the e-wave has an ellipsoidal wave front. The two wave fronts touch each other along the optic axis of the crystal. In a negative crystal the e-ray velocity is greater than the o-ray velocity and hence the ellipsoidal surface lies outside the spherical surface.

Let us consider that parallel beam of light CD falls *normally* on the surface of the negative crystal. As soon as the parallel beam strikes the crystal boundary, each point on the wave front becomes a source of secondary disturbance. The points A and B are chosen for the purpose of illustration. According to Huygens principle the points A and B produce elliptical and spherical wavelets. The position of the two wave fronts after a lapse of 't' seconds can be determined as follows. A circle of radius '$\upsilon_0 t$' is drawn taking A as centre. Similarly, a circle of the same radius is drawn with B as centre. The ellipsoidal wave front can be drawn if the major and minor axes are known. In a negative crystal the major axis is equal to '$2\upsilon_e t$' and minor axis is '$2\upsilon_e t$'. The circle and the ellipse touch each other, as required, along the optic axis. If we now draw the common tangents to the secondary wavelets, they represent the plane wave fronts corresponding to the two rays. Thus KL is the tangent to the spherical wavelets and EF is the tangent to the ellipsoidal wave fronts. If now we join the point of origin of wavelets to the points of tangency, the direction of propagation of the rays will be known. Thus, if we join A to K and B to L, we obtain the direction of propagation of o-ray. Similarly, the lines AE and BF show the direction of propagation of e-ray. We find that in this case the o-ray and e-ray travel along different directions with different velocities.

Case 2(a) : Optic axis in the plane of incidence and parallel to the refracting edge

Let MNN' M' represent the refracting face of the calcite (negative) crystal. Let CD represent the monochromatic plane wavefront incident normal to the crystal surface. Let the optic axis be *parallel to the refracting edge MN and lie in the plane of incidence.* Fig. 5(a). According to Huygens the phenomenon of double refraction involves propagation of light waves of two types. The wave known as o-wave has a spherical wavefront whereas the e-wave has an ellipsoidal wavefront. The two wavefronts touch each other along the optic axis of the crystal. In a negative crystal the e-ray velocity is greater than the o-ray velocity and hence the ellipsoidal surface lies outside the spherical surface.

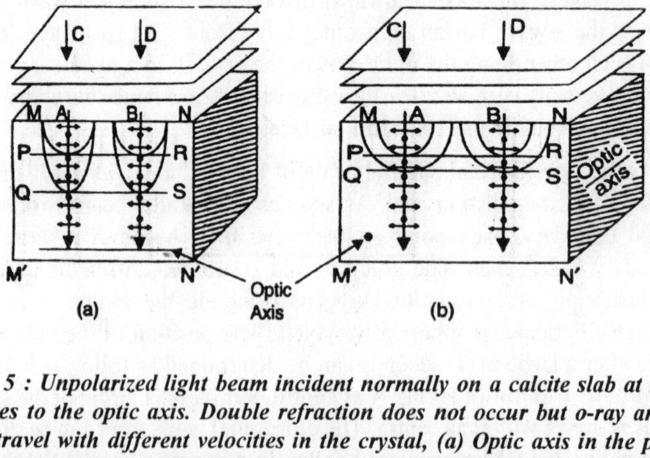

Fig. 5 : Unpolarized light beam incident normally on a calcite slab at right angles to the optic axis. Double refraction does not occur but o-ray and e-ray travel with different velocities in the crystal, (a) Optic axis in the plane of incidence and parallel to the crystal surface; (b) Optic axis perpendicular to the plane of incidence and parallel to the crystal surface.

Let us consider that parallel beam of light CD falls *normally* on the surface of the negative crystal. As soon as the parallel beam strikes the crystal boundary, each point on the wavefront becomes a source of secondary disturbance. The points A and B are chosen for the purpose of illustration. According to Huygens' principle the points A and B produce elliptical and spherical wavelets. The two wavelets touch each other along AM and BN, which is the direction of optic axis, as required.

The wave fronts at any instant can be drawn as tangential surfaces to the two wavelets. Note that the two wave fronts here are parallel to each other. The position of the two wavefronts after a lapse of 't' seconds can be determined as follows. A circle of radius '$\upsilon_0 t$' is drawn taking A as centre. Similarly, a circle of the same radius is drawn with B as centre. The ellipsoidal wavefront can be drawn if the major and minor axes are known. In a negative crystal the major axis is equal to '$2\upsilon_e t$' and minor axis is '$2\upsilon_0$'. PR is the tangent to the spherical wavelet and therefore represents the refracted wavefront for the o-ray. Similarly, QS represents the refracted wavefront for e-ray. If now we join the point' of origin of wavelets to the points of tangency, the direction of propagation of the rays will be known. Thus, if we join A to P and B to R, we obtain the direction of propagation of o-ray. By joining A to Q and B to S we get the direction of propagation of e-ray. We find that in this case the o- ray and e-ray travel along the same direction but have different velocities.

Case 2(b): Optic axis perpendicular to the plane of incidence and parallel to the refracting edge

Let MN be the refracting edge of the calcite (negative) crystal. Let CD represent the monochromatic plane wave front incident normal to the crystal surface. The optic axis is *parallel to the refracting edge but is perpendicular to the plane of incidence.* According to Huygens the phenomenon of double refraction involves two types of propagation of light waves. The wave known as o-wave has a spherical wave front whereas the e-wave has an ellipsoidal wave front. The two wave fronts touch each other along the optic axis of the crystal. In a negative crystal the e-ray velocity is greater than the o-ray velocity and hence the ellipsoidal surface lies outside the spherical surface.

Let us consider that parallel beam of light CD falls *normally* on the surface of the negative crystal. As soon as the parallel beam strikes the crystal boundary, each point on the wave front becomes a source of secondary disturbance. The points A and B are chosen for the purpose of illustration. According to Huygens principle the points A and B produce elliptical and spherical wavelets. Since the spherical and ellipsoidal wavelets touch each other along a line perpendicular to the plane of incidence, the wavelets appear spherical in the plane of incidence. The position of the two wave fronts after a lapse of 't' seconds can be determined as follows.

A circle of radius '$\upsilon_0 t$' is drawn taking A as centre. Similarly, a circle of the same radius is drawn with B as centre. The ellipsoidal wave front can be drawn if the major and minor axes are known. In a negative crystal the major axis is equal to '$2\upsilon_e$ t' and minor axis is '2'υ_0t'. The circle and the-ellipse touch each other, as required, along the optic axis, which is perpendicular to the surface MN. If we now draw the common tangents to the secondary wavelets, they represent the plane wavefronts corresponding to the two rays. Thus PR is the common tangent to the spherical wavelets and therefore represents the refracted wave front for the o-ray. Similarly, QS represents the refracted wave front for e-ray. If now we join the point of origin of wavelets to the points of tangency, the direction of propagation of the rays will be known. Thus, if we join A to P and B to R, we obtain the direction of propagation of o-ray. By joining A to Q and B to S we get the direction of propagation of e-ray. We find that in this case the o- ray and e-ray travel along the same direction but have different velocities.

Case 3 : Optic axis perpendicular to the refracting edge and lying in the plane of incidence

Let MNN' M' represent the calcite (negative) crystal. Let CD represent the monochromatic plane wave front incident normal to the crystal surface. The optic axis is *perpendicular to the refracting edge and lies in the plane of incidence.* According to Huygens the phenomenon of double refraction involves two types of propagation of light waves. The wave known as o-wave has a spherical wave front whereas the e-wave has an ellipsoidal wave front. The two wave fronts touch each other along the optic axis of the crystal. In a negative crystal the e-ray velocity is greater than the o-ray velocity and hence the ellipsoidal surface lies outside the spherical surface.

Let us consider that parallel beam of light CD falls *normally* on the surface of the negative crystal. As soon as the parallel beam strikes the crystal boundary, each point on the wavefront becomes a new point source of light. The points A and B are chosen for the purpose of illustration. According to Huygens' principle the points A and B produce elliptical and spherical wavelets. Since the two wavelets must touch each other along the optic axis, they touch at points P and Q (Fig. 6). Note that the section of the ellipsoid lies outside the section of the circle. The minor axis of the ellipse and the radius of the circle are equal. The position of the two wave fronts after a lapse of 't' seconds can be determined as follows. A circle of radius '$\upsilon_0 t$' is drawn taking A as centre. Similarly, a circle of the same radius is drawn with B as centre. The ellipsoidal wave front can be drawn if the major and minor axes are known. In a negative crystal the major axis is equal to '$2\upsilon_0 t$' and minor axis is '$2\upsilon_0 t$'. The circle and the ellipse touch each other, as required, along the optic axis, which is perpendicular to the refracting edge MN. If we now draw the common tangents to the secondary wave-lets, they represent the plane wavefronts corresponding to the two rays.

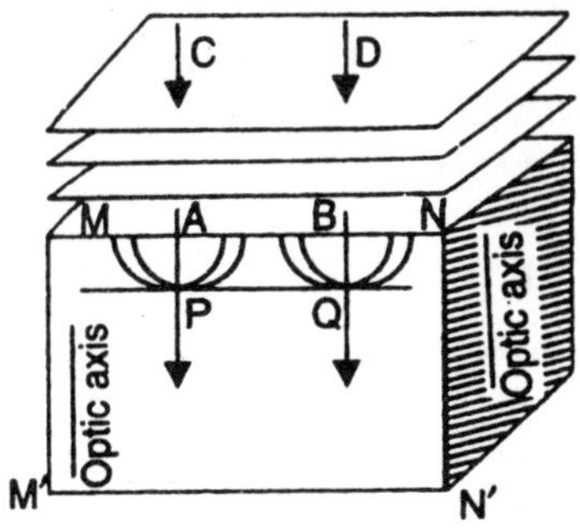

Fig. 6 : Unpolarized light beam incident normally on a calcite slab parallel to the optic axis. The o-ray and e-ray travel along the same velocity.

Thus PQ is the common tangent to the spherical wavelets as well as the ellipsoidal wave fronts. If now we join the point of origin of wavelets to the points of tangency, the direction of propagation of the rays will be known. Thus, if we join A to P and B to Q, we obtain the direction of propagation of o-ray as well as e-ray. We find that in this case the o- ray and e-ray travel along the same direction with the same velocity.

POLARIZATION BY SELECTIVE ABSORPTION

In 1815 Biot discovered that certain mineral crystals absorb light selectively. When natural light passes through a crystal such as tourmaline, it is split into two components, which are polarized in mutually perpendicular planes. The crystal strongly absorbs light that is polarized in a direction parallel to a particular plane in the crystal but freely transmits the light component polarized in a perpendicular direction. This difference in the absorption for the rays is known as *selective absorption* or *dichroism*. If the crystal is of proper thickness, one of the components is totally absorbed and the other component emerging from the crystal is linearly polarized. Selective absorption is illustrated. A good example of dichroic crystal is a tourmaline crystal.

The difference in absorption in different directions may be understood from the electron theory. When the frequency of the incident light wave is close to the natural frequency of the electron cloud, the light waves are absorbed strongly. Crystals that exhibit selective absorption are *anisotropic*. The crystal splits the incident wave into two waves. If the light wave frequency is close to the natural frequency of the electron cloud, then the component having its vibrations perpendicular to the principal plane of the crystal gets absorbed. The component with parallel vibrations is less absorbed and transmitted. The transmitted light is linearly polarised. However, the difficulty is that the crystals cannot be grown to sufficiently bigger

POLARIZATION BY DOUBLE REFRACTION

The double refraction phenomenon was discovered in 1669 by Erasmus Bartholinus during his studies on calcite ($CaCO_3$) crystals, which are known more as Iceland spar. When light is incident on a calcite crystal, it is split into two refracted rays (Fig. 7) differing in then-properties. The phenomenon of causing two refracted rays by a crystal is called *birefringence* or *double refraction*. The crystals are said to be birefringent.

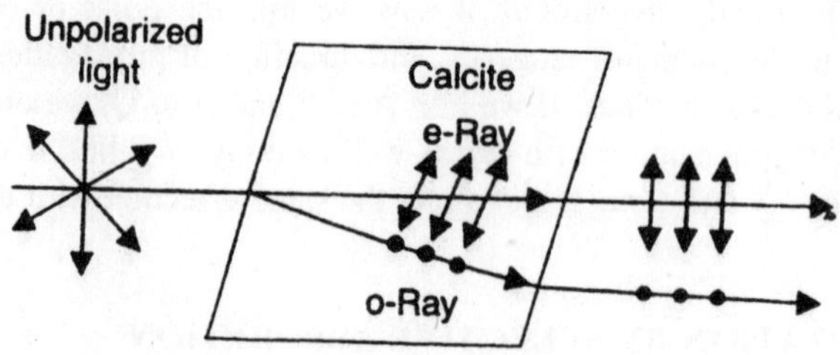

Fig. 7

The two rays produced in double refraction are linearly polarised in mutually perpendicular directions. One of the rays obeys Snell's law of refraction and hence is known as an ordinary ray or o-ray. The other ray does not obey Snell's law and is called an *extraordinary ray* or *e-ray*. Hence, the two rays can be distinguished from each other. If one of the rays is eliminated, the light transmitted by the crystal will be a linearly polarised light.

EXPERIMENTAL DETERMINATION OF PRINCIPAL REFRACTIVE INDICES

For determining the refractive index for the extraordinary ray a calcite crystal is cut in the form of a prism with the optic axis perpendicular to the refracting edge of the prism and perpendicular to the base BC. It can also be cut with the optic axis parallel to the refracting edge of the prism. The prism is placed on the spectrometer table and is adjusted for the minimum deviation position for the extraordinary rays. The angle of minimum deviation δ_m is determined and the principal refractive index for the extraordinary ray is calculated.

$$\mu_e = \frac{\left(\dfrac{A+\delta_m}{2}\right)}{\sin\dfrac{A}{2}} \qquad ...(1)$$

For a given wavelength, the ordinary and the extraordinary rays are separated while passing through the prism. Therefore, the angle of minimum deviation δ_m' for the ordinary ray can be measured and thus its refractive index can be calculated.

$$\mu_0 = \frac{\sin\left(\dfrac{A+\delta'_m}{2}\right)}{\sin\dfrac{A}{2}} \qquad ...(2)$$

PHASE DIFFERENCE BETWEEN e-RAY AND o-RAY

We have seen that natural light incident on the surface of an anisotropic crystal undergoes double refraction and produces two plane polarized waves. *Even when a plane polarized light wave is incident on a birefringent crystal such that its electric vector makes an angle with the optic axis, then the polarized wave splits into two polarized waves namely e-ray and o-ray*. Let us consider the particular case of a slice of a positive crystal where the optic axis is parallel to refracting face of the crystal. The two waves travel along the same direction in the crystal but with different velocities. As a result, when the waves emerge from the rear face of the crystal, an optical path difference would have developed between them. The optical path difference can be calculated as follows:

Let d be the thickness of the crystal.

The optical path for o - ray within the crystal $\Big\} = \mu_o d$

The optical path for e - ray within the crystal $\Big\} = \mu_e d$

The optical path difference between e - ray and o - ray $\Big\} \Delta = (\mu_e - \mu_o)d$ ...(1)

Consequently, a phase difference arises between the two waves. It is given by

$$\delta = \frac{2\pi}{\lambda}(\mu_e - \mu_o)d \qquad ...(2)$$

As the two component waves are derived from the same incident wave, the two waves are in phase at the front face and have emerged from the crystal with a constant phase difference and hence it may be expected that the waves are in a position to interfere with each other. However, as the planes of polarization of o-ray and e-ray are perpendicular to each other, interference cannot take place between e- and o-rays. The waves instead combine with each other to give an elliptically polarized wave.

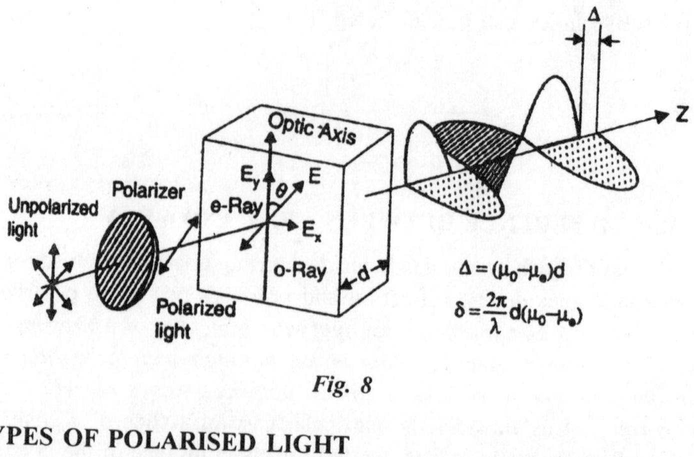

Fig. 8

TYPES OF POLARISED LIGHT

We can now sum up the various types of polarised light as follows.

(i) *Unpolarized light*, which consists of sequence of wave trains, all oriented at random. It is consi-dered as the resultant of two optical vector components, which are incoherent.

(ii) *Linearly polarised light*, which can be regarded as a resultant of two coherent linearly polarised waves.

(ii) *Partially polarised light*, which is a mixture of linearly polarised light and unpolarised light. Partially polarized light is represented as shown in Fig. 9.

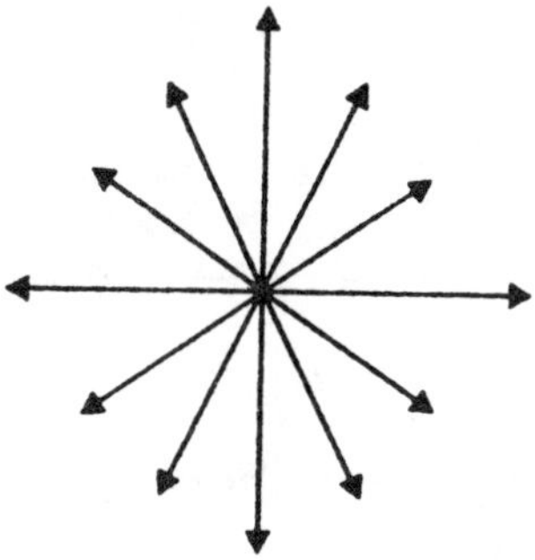

Fig. 9 : Schematic representation of partially polarized light.

If we imagine that we are looking at the light wave advancing towards us, we would observe that the tip of the E vector traces an ellipse. If we look from sides, we would find that the tip of E sweeps a flattened helix in space. When we are looking back towards the source, if the rotation of E vector occurs clockwise, it is said to be a right-*elliptically*-polarised wave. If it rotates anticlockwise, as we look back toward the source it is said to be a *left-elliptically polarised* wave.

(iv) *Circularly polarised light*, which is the resultant of two coherent waves having same amplitudes and a constant phase difference of 90°. A light wave is said to be *circularly polarised*, if the magnitude of the electric vector E stays constant but the vector rotates about the direction of propagation such that it goes on sweeping a circular helix in space (Fig. 10). If we imagine that the wave is advancing toward our eyes, we would find that the tip of the E vector of the wave traces a circle. If the rotation of the tip of E is clockwise, as seen by an observer looking back towards the source, then the wave is said to be right-circularly polarised. If the tip of E rotates anticlockwise, as seen by an observer looking back toward the source, the wave is said to be left-circularly polarised.

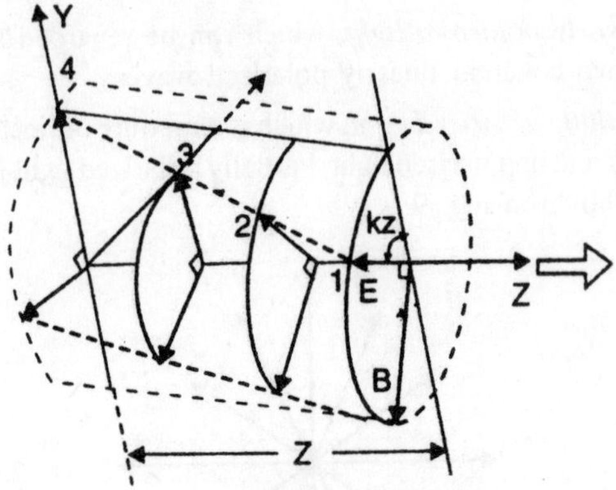

Fig. 10

ELLIPTICALLY POLARIZED LIGHT PRODUCTION

A quarter wave plate and a polarizer are the optical devices necessary to produce elliptically polarized light from unpolarized light. Unpolarized light is first converted to plane polarized light by allowing it to pass through a polarizer (a polaroid sheet or a Nicol prism). The plane polarized light is then made incident on a quarter wave plate (Fig. 11). The quarter wave plate or the polarizer is rotated such that the electric vector E of plane polarized light wave makes an angle θ ($\neq 45°$) with the optic axis of the quarter wave plate. The incident ray divides into o-ray and e-ray of amplitudes E sin θ and E cos θ. The rays travel along the same direction in the crystal with different velocities. The two rays are polarized in orthogonal planes. They are in phase at the front face but progressively get out of phase as they travel through the crystal. When they emerge out of the crystal they will have a path

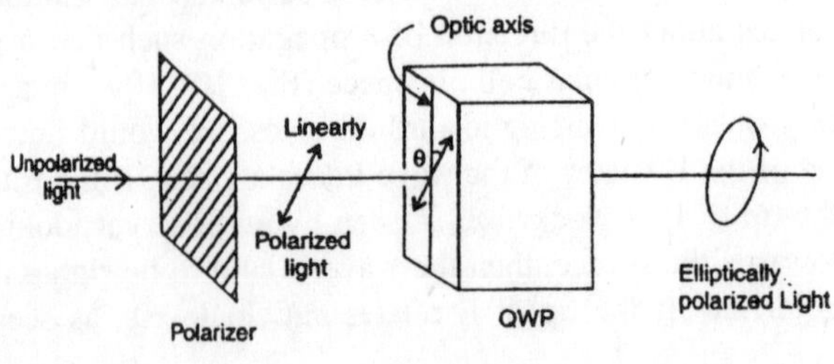

Fig. 11

difference of λ/4 or a phase difference of 90°. When they combine, they produce elliptically polarized light.

Detection of Elliptically Polarized Light

The light beam is allowed to pass through an analyser (a polaroid sheet or a Nicol prism). If on rotating the analysing polaroid sheet or Nicol, the intensity of the emerging beam varies from a maximum to a minimum value, but is never zero, then the incident light is elliptically polarized. A similar result would be obtained if the incident light is partially polarized. The two cases may be distinguished by inserting a quarter wave plate in the path of light before it falls on the analyser. If the original light is elliptically polarized, it may be considered as resultant of two coherent plane polarized waves, that is e-ray and o-ray, which are out of phase by 90°. If the light passes through the quarter wave plate, an additional phase difference of 90° is introduced between the e-ray and o-ray. Therefore, the total phase difference becomes 180° between the e-ray and o-ray. On emerging from the quarter plate, the e-and o-rays combine to produce plane polarized light. If the light coming out of quarter wave plate is examined with an analyzer, light will be extinguished twice in one full rotation of the polarizer as shown in Fig. 12.

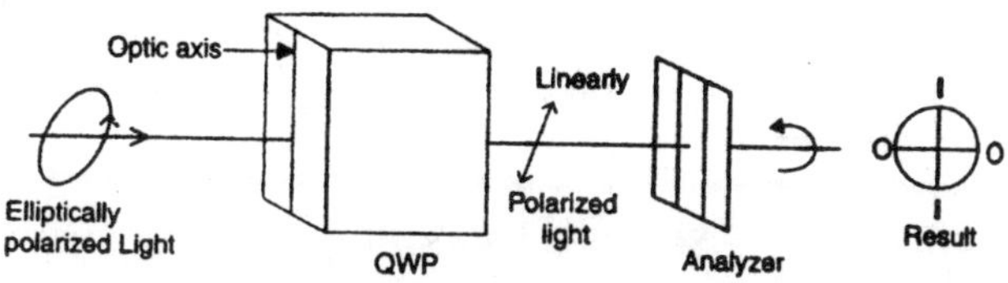

Fig. 12

CIRCUIARLY POLARIZED LIGHT PRODUCTION

A quarter wave plate and a polarizer are the optical devices required for producing circularly polarized light from unpolarized light.

Unpolarized light is first converted to plane polarized light by allowing it to pass through a polarizer (a polaroid sheet or a Nicol prism). Plane polarized light is then made to be incident on a quarter wave plate. The polarizer and the quarter wave plate are rotated such that the electric vector E of the plane-polarized wave makes an angle of 45° with the optic axis of the quarter wave plate. The plane polarized wave incident on the quarter wave plate splits into two rays, o-ray and e-ray of equal amplitude

($E_1 \cos 45° = E_2 \sin 45°$). The two rays travel in the same direction inside the crystal but with different velocities. The two rays are in phase at the front face of the crystal but progressively get out of phase as they travel through the crystal. As they emerge from the rear face of the crystal, they will have a path difference of $\lambda/4$ or phase difference of 90°. The two rays are linearly polarized in mutually perpendicular directions. When they combine, they produce circularly polarized light.

DETRACTION OF CIRCULARLY POLARIZED LIGHT

The light beam is allowed to pass through an analyser (a polaroid sheet or a Nicol prism). If on rotating the analysing polaroid sheet or Nicol, the intensity of the emerging beam remains uniform, then the incident light is circularly polarized. A similar result would be obtained if the incident light is ordinary unpolarized light. The two cases may be distinguished by inserting a quarter wave plate in the path of light before it falls on the analyser.

If the given light is circularly polarized, it may be considered as resultant of two coherent plane polarized waves, that is e-ray and o-ray, which are out of phase by 90°. If the light passes through the quarter wave plate, an additional phase difference of 90° is introduced between the e-ray and o-ray. Therefore, the total phase difference becomes 180° between the e-ray and o-ray. On emerging from the quarter plate, the e- and o-rays combine to produce plane polarized light.

SOLVED EXAMPLES

Example 1:

Unpolarized light falls on two polarizing sheets placed one on top of the other. What must be the angle between the characteristic of the sheets if the intensity of the transmitted light is one third intensity of the incident beam?

Solution:

Intensity of the light transmitted through the first polarizer $I_1 = I_o/2$, where I_o is the intensity of the incident unpolarized light.

Intensity of the light transmitted through the second polarizer is $I_2 = I_1 \cos^2\theta$ where θ is the angle between the characteristic direction of the polarizer sheets.

But $I_2 = I_o/3$ (given)

$\therefore \quad I_2 = I_1 \cos^2\theta = I_o \cos^2\theta/2 = I_o/3$

$\therefore \quad \cos^2\theta = 2/3$

$\therefore \quad \cos\theta = 0.8165$

$\therefore \quad \theta = 35.3^\circ$

Example 2:

If the plane of vibration of the incident beam makes an angle of 30° with the optic axis, compare the intensities of extraordinary and ordinary light.

Solution:

Intensity of the extraordinary ray $I_e = E^2 \cos^2\theta$

Intensity of the ordinary ray $I_o = E^2 \sin^2\theta$

$$\frac{I_e}{I_o} = \frac{\cos^2\theta}{\sin^2\theta}$$

$$= \frac{\cos^2 30^\circ}{\sin^2 30^\circ} = 3$$

Example 3:

A plane-polarized light is incident perpendicularly on a quartz plate cut with faces parallel to optic axis. Find the thickness of quartz plate, which introduces phase difference of 60° between e- and o-rays.

Solution:

The path difference between the waves is given by $D = (\mu_e - \mu_o)d$

The phase difference between the waves is given by $d = \dfrac{2\pi}{\lambda}\Delta$

$$D = \frac{60^\circ}{360^\circ}\cdot\lambda = \frac{\lambda}{6}$$

$$\therefore \quad d = \frac{\lambda}{6(\mu_e - \mu_o)}$$

$$d = \frac{5400\dot{A}}{6(1.553 - 1.544)}$$

$$= \frac{0.54}{0.054}\ \mu m = 10\mu m.$$

Example 4:

Calculate the thickness of double refracting plate capable of producing a path difference of $\lambda/4$ between extraordinary and ordinary waves. Given: λ = 5890Å,

$\mu_o = 1.53,\ \mu_e = 1.54.$

Solution:

$$d = \frac{\lambda}{4(\mu_e - \mu_o)} = \frac{5890\text{Å}}{4(1.54 - 1.53)}$$

$$= \frac{0.589}{0.04}\ \mu m = 14.7\mu m$$

Example 5:

Calculate the least thickness of a calcite plate which would convert plane polarized light into circularly polarized light. Given μ_o= 1.658, μ_e = 1.486 and wavelength of light is 5890Å.

Solution:

Plane polarized light gets converted into circularly polarized light by a suitably oriented quarter wave plate. Its thickness is given by

$$\therefore \quad d = \frac{\lambda}{4(\mu_o - \mu_e)} = \frac{5890}{4(1.658 - 1.486)}$$

or $\quad d = 0.856\mu m.$

Example 6:

For calcite, μ_o = 1.658, μ_E = 1.486 for sodium yellow. Calculate the thickness of the thinnest QW plate of calcite for sodium yellow.

Solution:

$$d = \frac{5893 \times 10^{-8} \text{cm}}{4(1.658 - 1.486)} \times \frac{5893}{4 \times .172} \text{x} 10^{-8} = 0.86 \times 10^{-4} \text{cm}$$

(Obviously such thin plate will have to be supported between glass plates).

Example 7:

A quarter-wave plate is meant for λ_0 = 5.893 × 10^{-5} cm. What phase retardation ϕ will it show for λ = 4.358 × 10^{-5} cm? (Neglect changes of μ_0 and μ_E with λ).

Solution:

Assuming that QW plate is of the smallest thickness,

$$|\mu_0 - \mu_E|\, d = \frac{1}{4} \times 5.893 \times 10^{-5} \text{ cm}$$

Phase retardation for $\lambda = 4.358 \times 10^{-5}$ cm

$$\phi = \frac{2\pi}{4.358\times10^{-5}} \times \frac{5.829\times10^{-5}}{4} = 0.67\pi \text{ radian.}$$

(Thus a QW plate for yellow is *not* a QW for other colours).

Example 8:

Plane polarized light passes through a calcite plate with its optic axis parallel to the faces. Calculate the least thickness of the plate for which the emergent beam will be plane polarized. Given $\mu_o = 1.6584$, $\mu_e = 1.4864$ *and wavelength of light is 5000Å.*

Solution:

When a plane-polarized beam is incident on a half wave plate, the emergent beam will also be plane polarized but the plane of polarization undergoes a rotation through an angle 2θ.

$$d = \frac{\lambda}{2(\mu_o - \mu_e)} = \frac{5\times10^{-7}\,\text{m}}{2(1.6584 - 1.4864)} = 1.45\ \mu\text{m}$$

Example 9:

The rotation in the plane in a certain substance is 10°/cm. Calculate the difference between the refractive indices for right and left circularly polarized light in the substance. Given λ = 5893Å.

Solution:

$$d = \frac{2\pi}{\lambda}[\mu_R - \mu_L]d \quad \therefore \quad [\mu_R - \mu_L] = \frac{\delta\lambda}{2\pi d}$$

It is given that $\dfrac{\delta}{d} = 10° = \dfrac{10\times2\pi}{360°} = \dfrac{\pi}{18}$ radian/cm

and $\lambda = 5893\text{Å} = 5893 \times 10^{-8}$ cm.

$$\mu_R - \mu_L = \frac{\pi}{18} \cdot \frac{5893\times10^{-8}\,\text{cm}}{2\pi}$$
$$= 1.6 \times 10^{-6}.$$

Example 10:

A 200 mm long tube containing 48 cm³ of sugar solution produces an optical rotation of 11° when placed in a saccharimeter. It the specific rotation of sugar solution is 66°, calculate the quantity of sugar contained in the tube in the form of a solution.

Solution:

It is given that $\theta = 11°$,

$$l = 200 \text{ mm} = 20 \text{ cm},$$

$$S = 66°, \text{ and } V = 48 \text{ cm}^3.$$

$$C = \frac{10\theta}{lS} = \frac{10 \times 11°}{20\text{cm} \times 66°} = 0.0833 \text{ g/cm}^3$$

Mass of sugar in solution

$$M = CV = 0.0833 \text{ g/cm}^3 \times 48 \text{ cm}^3 = 4 \text{ grams.}$$

Example 11:

20 gm of cane-sugar is dissolved in water to make 50 cc of solution. A 20 cm length of this solution causes + 53°30' optical rotation. Calculate the α.

Solution:

$$\theta = 53.5°, \ c = 20/50 = 0.40 \text{ gm/cc}, \ l = 2 \text{ decimeters.}$$

$$\therefore \quad \alpha = \frac{\theta}{cl} = \frac{107}{2} \times \frac{1}{0.40}$$

$$= 66.9 \frac{\text{deg}}{\text{dm}} \cdot \left(\frac{\text{gm}}{\text{cc}}\right)^{-1}.$$

Example 12:

If 20 cm length of a certain solution causes right-handed rotation 38°, and 30 and path of another solution causes left-handed rotation 24°, what optical rotation will be caused by 30 cm length of a mixture of the above solutions in volume ratio 1:2? (The solutions are not chemically reactive).

Solution:

The 30 cm length of the mixture may be treated as 10 cm of the first solution and 20 cm of the second (ratio 1 : 2).

∴ optical rotation by first is 38° × 10/20 = 19° (+)

optical rotation by second is 24° × 20/30 = 16° (–)

∴ Total rotation = 19° – 16° = 3° (+)

Example 13:

Quartz has specific rotation 21.7° per mm for l = 5893Å. A plane polarized beam is passed through a dextro-rotatory quartz plate of thickness 1.92 mm. Deduce the optical rotation.

Solution:

$$\theta = \alpha l = +21.7\ \frac{\text{deg}}{\text{mm}} \times 1.92\ \text{mm}$$

$$= +41.7\ \text{deg.}$$

Example 14:

Calculate the E.M.F. of the cell at 18°C

$$\underset{1\ \text{atm.}}{\text{Pt}\ H_2}\left|N/10\ \text{HCl}\ ||\ N\ \text{KOH}\right|\underset{1\ \text{atm.}}{\text{Pt}\ H_2}$$

Assume N/10 HCl as 90% dissociated and N KOH as 75% dissociated. Ionic product of water is 10^{-14}. Neglect the liquid-liquid potential.

Solution:

$$E = \frac{2.303\ RT}{F}\log\frac{C_2}{C_1} = 0.058\log\frac{C_2}{C_1}$$

Here C_1 = concentration of H^+ in N/10 HCl = 0.1 × 0.9 = 0.09

and C_2 = concentration of H^+ in N – KOH

$$= \frac{C_{H^+} \times C_{OH^-}}{C_{OH^-}} = \frac{K_w}{C_{OH^-}} = \frac{10^{-14}}{1.0 \times 0.75} = \frac{10^{-14}}{0.75}$$

$$\therefore \quad E = 0.058\log\frac{10^{-14}}{0.75 \times 0.09} = 0.058\log\frac{10^{-14}}{0.75}$$

$$= 0.058 \times (-12.8293)$$

$$E = -0.747\ \text{volt.}$$

Example 15:

What is the E.M.F. at 25°C of the cell

$$H_2\ |\ 0.5\ N - HCl\ |\ 0.1\ N\ NaOH\ |\ H_2$$

if the hydrogen at each electrode is under 1 atmospheric pressure and if the diffusion potential is neglected ?

The degree of dissociation of 0.5 N – HCl is 87% and that of 0.1 N NaOH is 90%. The ionic product of water is 1.2×10^{-14} at 25°C.

Solution:

The E.M.F. of the cell is given by

$$E = \frac{2.303\ RT}{nF} \log \frac{C_2}{C_1} = \frac{0.0591}{1} \log \frac{C_2}{C_1}$$

where C_1 and C_2 are the concentrations of H^+ ions in 0.5N HCl and 0.1 N NaOH respectively.

Now $C_1 = 0.5 \times 0.87$

$= 0.435$ gm ion/litre.

$$C_2 = C_{H^+} \text{ in NaOH} = \frac{C_{H^+} \times C_{OH^-}}{C_{OH^-} \text{ in 0.1 NNaOH}}$$

$$C_2 = \frac{k_w}{C_{OH^-} \text{ in 0.1 N} - \text{NaOH}} = \frac{1.2 \times 10^{-14}}{0.1 \times 0.9}$$

$$[\because k_w = C_H^+ \times C_{OH}^-]$$

$$= \frac{1.2 \times 10^{-14}}{0.090} = 1.333 \times 10^{-13} \text{ gm ion/litre}$$

$$E = 0.0591\ 10g = \frac{1.333 \times 10^{-13}}{0.435} = 0.0591 \times (-12.5326)$$

$E = 0.472$ volt.

Example 16:

The concentration cell

Ag | AgCl(s), KCl (0.5M) | K_xHg | KCl (0.05N), AgCl(s) | Ag

has an E.M.F. of – 0.1074 volts. Formulate the corresponding cell which transference, for which the E.M.F. is found to be – 0.0537 volt. Find out the transport number of chloride ion.

Solution:

The construction of the cell with transference will be

Ag | AgCl(s), KCl (0.5N) | KCl (0.05N), AgCl(s) | Ag

Being reversible with respect to chloride ion, this cell has an E.M.F. which is given by :

$$E_t = 2t + \frac{RT}{F} \log_e \frac{a_1}{a_2}$$

The E.M.F. of the given cell is : $E = \frac{2RT}{F} \log_e \frac{a_1}{a_2}$

Therefore $t^+ = \frac{E_t}{E} = \frac{-0.05357}{0.1074} = 0.498$

Transport number of chloride ion = 1 – 0.498 = 0.502.

Example 17:

Find the stability constant of the complex ion $[Cd(CN)_4]^{-2}$ at 25° given that the complex ion and KCN are ionised to 90% fro concentration upto 2N. The E.M.F. of the cell

Cd | 0.05N Cd $(NO)_2$:

NH_4O_3 : 0.01N $Cd(NO_3)_2$ | Cd

(in 2N KCN)

is + 0.578 volt at 25°C.

Solution:

The E.M.F. of the cell is given by

$$E = \frac{RT}{2F} \log_e \frac{C_1}{C_2} = \frac{RT}{2F} \log_e \frac{0.1 \times 0.9}{x}$$

where x = concentration of Cd^{2+} ions in the left hand side compartment.

Therefore, at 25°C, $0.578 = \frac{0.0591}{2} \log_e \frac{0.09}{x}$

$x = 3.937 \times 10^{-10}$ gm equiv./litre

The instability constant of the complex cyanide ion (K_t) is

$$K_t = \frac{[Cd^{2+}][CN^-]^4}{[Cd(CN)_4{}^{2-}]} = \frac{3.937 \times 10^{-22} \times 2^4}{0.05 \times 0.9}$$

$= 1.4 \times 10^{-17}$.

EXERCISES

1. Calculate the polarizing angle for an air-diamond interface when light is incident (i) from air, (ii) from diamond. (Given n of diamond = 1.8,
2. A biaxial crystal has wave-velocities V_0 and V_E. The optic axis is designated as z-axis. For a wave with wave-normal atong the Jr-axis, deduce the velocities and directions ofE vectors for the two polarized waves possible.
3. State the laws of rotatory polarisation. Give Fresnel's hypothesis for rotatory polarisation and derive a formula for the rotation of quartz. Give the experimental verification of this formula.
4. A transparent plate is given. Using two Nicol prisms how would you find whether the given plate is a quarter wave plate, a half wave plate or a simple glass plate.
5. On introducing a polarimeter tube 25 cm long and containing sugar solution of unknown strength, it is found that the plane-of polarisation is rotated through 10°. Find the strength of the sugar solution in g/cm^3. Given that the specific rotation of sugar solution is 60° per decimetre per unit concentration.
6. How will you orient the polarizer and analyser so that a beam of natural light is reduced to (i) 0.5 (ii) 0.25 (iii) 0.75 and (iv) 0.125 of its original intensity?
7. Sunlight reflected from still water in a lake is observed throu) Polaroid. When the polaroid is rotated' in its own plane the inte changes. Explain.
8. A solution of concentration 6.0 gm per 100 cm^3 used in a tube of length 33 cm causes 14° 30' rotation in the plane of polarisation of light of λ = 5500Å. Deduce the specific rotation. Also estimate the rotation it would cause for λ = 4500 Å.
9. 20 cm path of a solution containing 10 gm of material A per 100 cm^3 causes optical rotation + 25°, and 30 cm path of a solution containing 30 gm of material B per 100 cm^3 causes optical rotation –15.° Deduce the optical rotation due to 40 cm path of a solution containing 8 gm of A and 6 gm of B in 50 cc of the solution. State any assumption made.
10. An unpolarized beam of light is passed through a double image prism. Then one beam is passed through a quartz plate to rotate

the plane of polarization by 90°, and this is then superposed with the other beam. Discuss if interference fringes would occur. Howwould things change if the starting beam itself were plane polarized?

11. Outline Fresnel's explanation of optical rotation. Discuss :

    (i) the dependence of rotation on λ,

    (ii) the experimental evidence in support of the assumptions.

12. In a Laurent half-shade device should the optic axis of the HW plate be parallel to the diameter? Could the portion covered by the simple glass plate be left just vacant? Where should the telescope of the polarimeter be focused and why?

13. In measuring concentration with a polarimeter, a sample tube of length 20.0 ± 0.1 cm is found to show an optical rotation + 32.6 ± 0.1 ± with λ = 5893 A. Tne spednc rotation is known for this A, tobe + 58.2 ± 0.1° per dm per gm/cc, Deduce the range of error in the concentration as obtained from these measurements.

14. The settine of a Laurent half-shade in a polarimeter was such that the optic axis of half-shade made 10° angle with the principal plane of the polarizer. Deduce the transmitted intensity and 'contrast for $\Delta\theta$ = 0.1.°' How would these change if the setting is changed to 5°?

15. If the Optic axis of a uniaxial crystal is taken as z-axis, write down the equation for the extraordinary wave-surface at unit time in terms of V_0 and V_E. For extraordinary ray in xz plane at angle 30° from x-ad deduce the (i) ray velocity, and (ii) the corresponding wave velocity and direction of the wave-normal.

16. Using Huygens' construction, show that a beam incident along the normal on a plate of biaxial crystal would split into two beams with a lateral separation proportional to, thickness of the plate. What other factors determine this separation?

17. A 'polarizer' is that which would convert ordinary unpolarized light to plane polarized light. The ideal polarizer is that which would give 50% of the original light as plane polarized one. Discuss this 50% business.

5

Photometry and Atom Laser

INTRODUCTION

In this Chapter we will be dealing with visual photometry only. The final measuring device is then the human eye. Therefore, we will also include a brief description of the eye and its functioning. The word 'Photometry' in its general meaning involves the measurement of radiant energy. However, out of the total electromagnetic spectrum the human eye is sensitive to only a small region of wavelengths and even within this region the visual sensation for equal radiant energy is not uniform. For this reason photometry in terms of *visual sensation* is very different from that based on measurement of radiant energy.

THE FUNDAMENTAL SI UNIT OF INTENSITY

A 'standard candle' was used early to define intensity. But the latest fundamental unit is *Candela defined as the luminous flux emitted per unit solid angle along normal to the surface by one-sixtieth cm2 of a black-body radiator kept at the temperature of solidification of platinum* (2,046 K). Thus 1cm^2 of this blackbody surface at 2,046 K will have intensity 60 cos θ cd in a direction Q from the normal. The brightness of a blackbody is the same in all directions, and its value for the 2,046 K surface would be 60×10^4 cd/m^2.

THE ULTRAVIOLET CATASTROPHE

In practice, a perfect black body is made by taking a hollow sphere (cavity) and drilling a small hole. The hole acts a perfect absorber. Light entering the cavity undergoes multiple reflections at the walls and gets trapped inside the cavity. Consequently, the hole appears perfectly dark. Conversely, when the cavity is heated, the radiation produced in the cavity comes out through the aperture and contains all the wavelengths.

Therefore, the hole acts as a perfect emitter. The spectral distribution of that radiation is a function of temperature alone and the material as such plays no role. Various efforts were made to calculate theoretically the frequency distribution of thermal radiation. At that period of time electron was discovered but the inner structure of the atom was unknown. The generally accepted model was that a blackbody was made up of a huge number of atoms and the atoms are regarded as small harmonic oscillators. The random thermal motion of atoms within the walls generates electromagnetic waves, which is the *thermal radiation* emitted from the walls of the cavity.

The radiation emitted by the atoms is reflected back and forth by the cavity walls to form a system of standing waves for each frequency present. There would be many modes of vibration present in the cavity space. Finally, when thermal equilibrium is attained, the average rate of emission of radiant energy by atomic oscillators in the walls equals the rate of absorption of radiant energy by the interior walls. According to Boltzmann's *principle of equipartition of energy*, each simple harmonic oscillator has an average thermal energy of 'kT' at thermal equilibrium. Rayleigh assumed that each of the standing waves ought to have energy 'kT and derived an expression for the energy density of radiation distribution within the cavity.

$$R(v) = \frac{8\pi v^2 kT}{c^3} \qquad ...(1)$$

This result is known as Rayleigh-Jeans law. The energy density calculated with the above formula agrees well with the experimental results at long wavelength end of the spectrum but goes toward infinity at the short wavelength end. It predicted that the intensity of thermal radiation should increase with square of frequency. It implied that the radiation emitted by a hot body should have a large portion of UV rays. This is contrary to our experience and violates the law of conservation of energy. This contradiction came to be known as ultraviolet catastrophe. The failure of Rayleigh-Jeans formula presented a crisis and gave the first indication of inadequacy of classical physics.

THE PHOTON

Max Planck introduced the concept of discontinuous emission and absorption of radiation by bodies but he treated the propagation of radiation through space as occurring in the form of continuous waves as demanded by electromagnetic theory. Einstein refined the Planck's hypothesis and invested the quantum with a clear and distinct identity. The energy quanta are named *photons*.

1. According to Einstein the quantization of energy, which is present in the emission and absorption processes, is retained as the energy propagates through space. A light beam is regarded as a stream of photons travelling with a velocity 'c'.
2. An electromagnetic wave having a frequency contains identical photons, each having an energy hv. The higher the frequency of the wave, the higher is the energy content of each photon. Thus, y-ray and .x-ray photons are more energetic compared to optical photons while the photons of r.f. frequencies are the feeblest.
3. The intensity of a monochromatic light beam / is related to the concentration of photons present in the beam.

$$I = Nhv \qquad ...(1)$$

4. When photons encounter matter, they impart all their energy to the particles of matter and vanish.

Two experiments, namely the photoelectric effect and Compton effect, have provided the convincing experimental evidence for the existence of photon.

VISUAL PHOTOMETRY : RELATED QUANTITIES

The basic measure of visual sensation created by the total radiation emanating from a source or incident on a receiver is called the luminous flux, expressed by the symbol ϕ, and measured in a unit *lumen* (lm). We talk of a given area S of a source giving out luminous flux ϕ lm in a specified solid angle Ω. We also talk of a screen of area S receiving luminous flux ϕ lm from one or many sources.

Out of the quantities defined, brightness (B) and, intensity (I) are direction dependent. In contrast, luminance is the integrated effect of B summed for all directions. Also, for a source of finite dimension, I is the integrated effect of B over the entire area of the source. Finally ϕ is the integrated effect of I over any given solid angle or say receiving area.

It may be noted that if solid angle is taken to be dimensionless, then there can be considerable mix-up in units. For this reason, we take the dimension of solid angle as $\left(L_y^2\ L_x^{-2}\right)$, where L_x and L_y are lengths along and transverse to any ray, respectively.

SPECTRUM AND SPECTRAL LINES

The next question we are concerned is about how the photons are emitted. In 1666 Newton discovered that different coloured rays of light were refracted at different angles when sunlight coming through a slit was passed through a prism. When a lens was kept in the path of the band of colours, a series of coloured images of the slit were found on a screen. Newton called these series of coloured slits a *spectrum*. In general, solid materials and gases at high pressure produce *continuous spectra*, whose investigations led to quantum hypothesis by Planck.

In contrast, vapours and gases at low pressure, where the atoms or molecules are far apart and do not significantly interact, emit *line spectra*. Each chemical element has unique spectral lines that are characteristic of the element

The set of all characteristic frequencies is designated as the *emission spectrum* of the substance. Some elements, such as hydrogen and neon, have relatively few lines; others, such as iron have many thousand lines. In the spectrum, the lines follow each other in an *orderly* fashion. A group of such lines is called a *series*. The study of line spectra led to the understanding of the structure of the atom and the mechanism of emission of photons by atoms.

In 1859, Gustav Kirchhoff summed up his observations on line spectra as follows:

1. Light from a hot body on passing through a prism gives a continuous spectrum with no lines.
2. The same light when passed through a cool gas yields continuous spectrum but with certain wavelengths of light removed.
3. If the light emitted by a hot gas is viewed through a prism, one observes a series of bright lines at specific wavelengths against an otherwise dark background.

The emission spectra of atoms, molecules, and solids differ considerably from each other. Atomic spectra lie mostly in the visible and ultraviolet regions and the lines are sufficiently spaced apart and hence atomic spectra are line spectra. Molecular spectra extend from the far infrared up to the ultraviolet and are composed of closely spaced lines which appear as bright bands in a spectroscope having low resolving power. Hence molecular spectra are called *band spectra*. It is already learnt that solids give continuous spectra.

The studies on atomic spectra led to the formulation of atomic models. In 1885 Johann Balmer found that for the hydrogen atom the wavelengths of the regular array of spectral lines are well described by a simple empirical formula,

$$\lambda = 3646\left[\frac{n^2}{n^2-4}\right] \text{Å} \quad n = 3,4,5.... \qquad ...(1)$$

which was later rearranged by J. Rydberg in the following form.

$$\lambda = R\left[\frac{1}{2^2} - \frac{1}{n^2}\right] \qquad ...(2)$$

where R is a constant and known as Rydberg constant and has the value

$$R = 1.0973732 \times 10^7 \text{ m}^{-1}. \qquad (3)$$

The reciprocal of wavelength I/A, is called the wave number.

Atomic Structure

J.J. Thomson discovered electron in 1897. He put forward the first ever model of the atom. Thomson represented it as a positively charged sphere with electrons immersed in it here and there. When unperturbed, both electrons and the positive charge were believed to be at rest. When in motion the electrons lose their energy in the form of radiation. Basing on the results of α-scattering experiments conducted by him, Rutherford proposed 'planetary model' of the atom. The atom consists of a relatively small positive nucleus at the centre and electrons orbiting well away from it. But an electron moving in an orbit around the nucleus is in a state of accelerated motion and according to Maxwell, any accelerated charge generates electromagnetic waves.

Consequently, the orbiting electron loses continually its energy until it approaches the nucleus and falls into it. This leads to the prediction that an atom exists for only about 10^{-8}s, but in fact atoms have remarkable stability. Secondly, if an electron is rotating around the nucleus, it radiates light whose frequency is the same as that of the rotation. As the electron spirals and falls onto the nucleus the light frequency continually increases and therefore the radiation given out by Rutherford atom must be continuous. However, the atomic spectra were in fact a set of discrete lines. In 1913, Niels Bohr removed the deficiencies of Rutherford model by incorporating some adhoc quantum hypotheses into it.

Bohr explained the origin of line spectra in general terms on the basis of two central ideas. One is the concept of photon and the other is the concept of energy levels of atoms. Bohr combined these two concepts nicely to explain how light is emitted and why the spectral lines are arranged in an orderly fashion.

The Frank-Hertz Experiment

In 1914, James Frank and Gustav Hertz found direct experimental evidence for the existence of atomic energy levels. Frank and Hertz studied the motion of electrons through mercury vapour under the action of electric field. They found that when an electron has kinetic energy 4.9 eV or greater, the vapour emitted UV light of wavelength 0.25 mm. Suppose mercury atoms have an energy level 4.9 eV above the lowest energy level. An atom can be raised to this level by collision with an electron; it later drops back to the lowest energy level by emitting a photon. According to eqn. (18) the wavelength of the photon should be

$$\lambda = \frac{hc}{E} = \frac{(4.136 \times 10^{-15}\,\text{eV.s})\,(3 \times 10^{8}\,\text{m/s})}{4.9\,\text{eV}} = 0.25\,\mu\text{m}$$

This is equal to the measured wavelength, confirming the existence of this energy level of the mercury atom. Similar experiments with other atoms yielded the same kind of evidence for atomic energy levels.

ATOMIC TRANSITIONS

The transition of atom from one energy state to another is accomplished by a transfer of energy. If energy is supplied to the system consisting of atoms, the atom is raised from a lower energy state E_1 to a higher excited state, E_2. Such a transition, $E_1 \rightarrow E_2$ is called absorption. The transition from a lower state to an excited state can occur only if the difference in energy is exactly equal to the photon energy hv. The atom does not stay in an excited state indefinitely. It usually returns on its own to the lower state after about 10 ns by emitting a photon. The downward transition $E_2 \rightarrow E_1$ is called *emission.* During the emission process, as the atom returns from a higher energy state E_2 to a lower energy state E_1, it emits a quantum of energy, hv. If the photon energy is $hv = \frac{hc}{\lambda}$, then conservation of energy gives

$$hv = \frac{hc}{\lambda} = E_2 - E_1 \qquad ...(1)$$

The jumps that electrons can make from one level to another. The location of these transitions, or *emission lines*, can be shown on a wavelength graph. The ensuing pattern corresponds exactly to that recorded by the spectrographic plates. Thus, *position* is one of the important properties of spectral lines. The average time spent by the atom in an excited level is called the *lifetime* of the level. The lifetime is usually of the order of 10^{-8}s. Besides the short-lived excited states, there are states with average lifetimes greater than 10 milliseconds or even as long as several seconds. They are called *metastable states.*

Limitations of Bohr Theory

Bohr's theory employed a semi-classical model where quantum postulates are introduced into a mechanical model. But it explained with remarkable success the origin of line spectra in case of hydrogen atom, which is a one-electron atom. However the theory could not give correct predictions for two electron atoms. The Bohr hypothesis established the relation of wavelengths to energy levels, but it did not provide any general principles for predicting the energy levels of a particular atom. It also did not explain why angular momentum is quantized and why and when atoms make transitions. Further, it cannot determine what the intensity of the spectral line will be. Inspite of such limitations, Bohr's theory contributed immensely to the understanding of atomic structure. A more general understanding of atomic structure and energy levels is provided by the quantum mechanics.

DE BROGUE HYPOTHESIS

We learnt earlier that a light beam consists of electromagnetic waves and in this chapter, we have learnt that it consists of a stream of photons. Thus, we find that light has dual nature and behaves as waves sometimes and as particles at other times.

As a photon travels with the velocity c, we can express its momentum as

$$p = \frac{E}{c} = \frac{hv}{c} = \frac{h}{\lambda} \qquad ...(1)$$

Thus, the wavelength and momentum p of a photon are related to each other through the expression

$$\lambda = \frac{h}{p} \qquad ...(2)$$

The quantities v and λ are wave properties and the quantities E and p are particle properties. The relations (1) and (2) demonstrate that the wave and particle natures of a photon are intimately tied up to each other.

In 1924, de Broglie suggested that the wave-particle dualism need not be a special feature of light. The relation (2) between the momentum and the wavelength of a photon must be a universal relation applying to photons and material particles alike. A particle of mass 'm' moving with a velocity υ carries a momentum

$$p = m\upsilon$$

and it must be associated with a wave of wavelength

$$\lambda = \frac{h}{p} = \frac{h}{m\upsilon} \qquad ...(3)$$

The waves associated with moving particles are called *matter waves or de Broglie waves*. The relation $\lambda = h/m\upsilon$ is known as de *Broglie equation* and the wavelength is called the *de Broglie wavelength.*

If the de Broglie hypothesis is valid, then the waves associated with matter should suffer diffraction. A beam of electrons diffracted by a crystal should show interference phenomena. Davisson and Germer obtained, in 1927, the first experimental evidence that gave convincing proof of the wave nature of matter.

De Broglie's Justification of Bohr's Postulate

One of the postulates that Bohr used in formulating a model of atom is that the angular momentum L of the electron revolving in a stationary orbit is quantized. Thus,

$$L = n\hbar \text{ —Bohr's, postulate} \qquad ...(4)$$

The above postulate of Bohr follows directly from the concept of matter waves. If a stretched string is fastened at both ends and is made to vibrate, standing waves are formed provided the length of the string is an integral number of half-wavelengths of the disturbance. If the string is formed into a circular loop, the condition for standing waves is that the circumference of the loop should be an integral number of whole wavelengths of the disturbance. Thus, if r is the radius of the circular loop,

$$2\pi r = n\lambda \qquad n = (1, 2, 3, ---) \qquad ...(5)$$

We may regard the stationary electron orbits in an atom to be analogous to the circular loop of string. We conclude that stationary electron wave pattern can form in the orbit if only an integral number of electron wavelengths fit into the orbit.

The above equation (5) can be applied for electron waves, taking λ as de Broglie wavelength of electron waves.

The de Broglie wavelength of electron wave is given by

$$\lambda = \frac{h}{m\upsilon}$$

where υ is the speed of the electron in the orbit. Using de Broglie wavelength into eqn. (31), we obtain

$$2\pi r = \frac{nh}{m\upsilon}$$

$$\therefore \qquad m\upsilon r = \frac{nh}{2\pi}$$

But the quantity '$m\upsilon r$' is the angular momentum, L, of electron in the orbit of radius r. Thus,

$$L = m\upsilon r$$

It, therefore, follows that

$$L = n\hbar \qquad ...(6)$$

which is precisely Bohr's postulate. De Broglie thus demonstrated that the quantization of angular momentum is a direct consequence of wave nature of electron.

Let us calculate the wavelength of the electron in the first orbit of hydrogen atom. The electron speed υ in the orbit is given by

$$\upsilon = \frac{e}{4\pi\varepsilon_o mr}$$

$$\therefore \qquad \lambda = \frac{h}{e}\sqrt{\frac{4\pi\varepsilon_o r}{m}} \qquad ...(7)$$

Taking $r = 5.3 \times 10^{-11}$ m, we get $\lambda = 3.3$Å. The circumference of the orbit is $2\pi r = 3.3$Å. Hence, the first orbit of the electron in a hydrogen atom corresponds to one complete electron wave joined on itself. The de Broglie hypothesis thus offered a new meaning to the principal quantum number "n". n is the number of de Broglie wavelengths that fit into the circumference of Bohr allowed orbits.

HEISENBERG UNCERTAINTY PRINCIPLE

The wave nature of moving particles leads to some inevitable consequences. Classically, the state of a particle is defined by its position and momentum. At each instant, the position and momentum of a classical particle can be measured to a very high accuracy. In case of a quantum particle, there are uncertainties associated with its location and momentum.

At about the same time that Davisson and Germer conducted the experiments of wavelike properties of electrons, Heisenberg put forth his *uncertainty principle*. He suggested that the product of uncertainty in the location of the quantum particle and the uncertainty in its momentum. $\Delta x \ \Delta p$, would always be of the order of Planck's constant h. Thus,

$$\Delta x \ . \ \Delta p + h \qquad ...(1)$$

This is known as Heisenberg uncertainty principle for position and momentum, which may be stated as follows:

"It is not possible to know simultaneously and with exactness both the position and the momentum of a microparticle".

This uncertainty principle expresses a fundamental limitation in nature that also limits the precision of our measurements.

The uncertainty principle asserts that it is physically impossible to know simultaneously the exact position ($\Delta x = 0$) and exact momentum ($\Delta p_x = 0$) of a microparticle. According to it, the more precisely we know the position of the particle, the less precise is our information about its momentum. The momentum of a particle cannot be precisely specified without our loss of knowledge of the position of the particle at that time. Similarly, a particle cannot be precisely localized in a particular direction without our loss of knowledge of momentum in that particular direction. We can at best specify that certain momentum of the particle is more probable than other or that the particle is more likely to be here than there. Thus, the uncertainly principle implies that we can never define the path of an atomic particle with the absolute precision indicated in classical mechanics. To describe the quantum particle the concept of *energy* becomes important since it is related to the *state* of the system rather than to its path.

In addition to the uncertainty relation between co-ordinates and momentum of a moving particle, there is an uncertainty relation between energy and time. Suppose that we want to determine the energy of a particle and the time at which the particle has such energy. If ΔE and Δt are the uncertainties in the values of these quantities, then

$$\Delta E \ . \ \Delta t \approx \hbar \qquad ...(2)$$

Thus, it is impossible to know simultaneously and with exactness the energy of a particle and the time at which it has that energy.

In view of the wavelike properties of particles and uncertainties in determining the physical parameters, a theory taking into consideration of the *probability distribution* of parameters such as position, momentum, energy in place of *precise values* is required. Efforts in this direction were done independently by Schrodinger and Heisenberg who used different approaches namely wave mechanics and matrix mechanics but which were found to be equivalent.

SCATTERING

When the electrical polarizability of molecule changes during its motion, light scattering takes place. This phenomenon also gives rise to light emission but does not involve electron transitions from a higher energy level to a lower energy level.

Raman Effect

In 1928, Sir C.V. Raman discovered that when a beam of monochromatic light was passed through an organic liquid such as benzene, the scattered light was found to contain a strong line of frequency v_o equal to the frequency of the incident light and a few weak lines on either side of the incident line $v' = v_o \pm v_M$. This is referred to as *Raman Scattering*. The scattering of light with change of frequency is known as *Raman effect*. The spectrum formed due to Raman effect is called *Raman spectrum* and the spectral lines obtained are called *Raman lines*.

In order to observe Raman effect, the incident light should be monochromatic and very intense. Raman scattering is always accompanied by Rayleigh scattering. Raman lines at frequencies less than that of the incident frequency ($v_o - v_M$) are known as Stokes lines and those with frequencies greater than that of the incident frequency ($v_o + v_M$) are known as *anti-Stokes* lines. Experiments show that Stokes lines are far more intense than anti-Stokes lines. Overall, however, the total radiation scattered at any but the incident frequency is extremely small, and sensitive apparatus is needed for its detection and study. If the frequency of the incident light is varied, the weak lines are once again observed on either side of the Rayleigh line with the same difference in frequency. It is evident that the frequency difference v' between the incident and the scattered light in Raman effect is determined by the nature of the scattering molecules and is independent of the frequency of incident light.

CLASSICAL EXPLANATION

When a sample of molecules is subjected to a beam of light of frequency v, the induced dipole undergoes oscillations of frequency v;

$$\mu = \alpha E = \alpha E_o \sin 2\pi vt \qquad ...(1)$$

If in addition, the molecule undergoes some internal motion, such as vibration or rotation, which changes the *polaribility* periodically, then the oscillating dipole will have superimposed upon it the vibrational or rotational oscillation. As an example, let us assume a vibration of frequency v_{vib}, which changes the polarizability. Then

$$\alpha = \alpha_o + \beta \sin 2\pi v_{vib} t \qquad ...(2)$$

where α_o is the equilibrium polarizability and p represents the rate of change of polarizability with the vibration. Using (2) into (1), we get

$$\mu = \alpha E = [\alpha_o + \beta \sin 2\pi v_{vib}\ t]\ E_o \sin 2\pi vt \qquad ...(3)$$

Expanding the above equation and using the trigonometric relation, we get

$$\mu = \alpha_o E_o \sin 2\pi vt + 1/2\ \beta E_o [\cos 2\pi (v - v_{vib}) t - \cos 2\pi (v + v_{vig}) t] \qquad ...(3)$$

It means that the oscillating dipole has frequency components $v \pm v_{vib}$, asrwell the exciting frequency v. These additional frequencies are responsible for producing the Raman lines. Thus, for Raman scattering, *a molecular rotation or vibration must cause some change in a component of molecular polarizability.*

Quantum Theory

The explanation of Raman scattering in terms of the quantum theory is very simple. When light is incident on a solid, liquid, or gas, the photons can be imagined to undergo collisions with molecules.

(i) If the collision is perfectly *elastic*, there will be no transfer of energy from the photon to the molecule or from the molecule to the photon. The photon is scattered without any change of energy. Therefore, the frequency of the scattered photon is the same as that of the incident photon. A detector placed to collect energy at right angles to an incident beam will thus receive photons of energy hv, that is the radiation of frequency V. This explains the Rayleigh line in the Raman spectrum.

(ii) However, it may happen that energy is exchanged between photon and molecule during the collision; such collisions are called *inelastic collisions*. The molecule can gain or lose amounts of energy only with the accordance with the quantum laws. That is its energy changes must be the difference in energy between two of its allowed states. That is to say, ΔE must represent a change in the vibrational and/or rotational energy of the molecule.

If the molecule gains energy ΔE (= $E_2 - E_1$) from the photon, it goes from a lower energy level E_1 to a higher energy level E_2, while the photon will be scattered with energy (hv – ΔE). Then, the scattered radiation will have a frequency (v – ΔE/h) which is less than that of the incident radiation. The resulting lines are Stokes lines located on the lower frequency side of the Raman spectrum.

Conversely, the molecule may be initially in an excited state. It may lose energy ΔE to the photon and go to a lower state after collision, while the photon is scattered with energy (hv + ΔE). Hence, the scattered radiation will have a frequency (v + ΔE/h). The resulting lines are anti-Stokes lines, which are situated on the higher frequency side of the Raman spectrum.

At ordinary temperatures, there are more molecules in the lower energy state. Therefore, transitions are more likely from lower energy state to upper energy state. Hence, energy is absorbed by the molecules from the photons. The reverse process of energy being given to photons by molecules is less likely. Therefore, Stokes lines are more intense than the anti-Stokes lines. The intensity of anti-Stokes lines increases with temperature, as a rise in temperature increases the relative population of molecules in the higher energy state.

WAVE FUNCTIONS

In quantum mechanics, various atomic parameters are described with the help of wave functions ψ. In general, these wave functions ψ are mathematical functions that depend on the variables necessary to specify the particular features of the particle. A wave function is a complex quantity and is not an *observable* quantity; hence it has no direct physical significance. For example, let the location of an electron be described by the wave function ψ (x, y, z, t).

This wave function by itself cannot indicate us the location of the electron. It is known that the intensity of a wave motion is proportional to the square of the amplitude of the wave. Therefore, the square of the absolute value $|\psi|^2$ of the wave function gives where the intensity of the field $|\psi|^2$ is large and thus the regions of space where the particle is more likely to be found at time ψ. The function y is, therefore, called the *probability amplitude*. Thus,

Probability, P, of finding the particle in an infinitesimal volume dV(= dx dy dz) is proportional to $|\psi(x, y, z)|^2$ dx dy dz at time t.

$$\text{or} \quad P \propto |\psi(x, y, z)|^2 \, dV \qquad ...(1)$$

The probability is a real quantity.

The probability of finding the particle between the limits x = a and x = b, y = g and y = h and z = p and z = q, is then

$$P \propto \int_a^b \int_g^g \int_p^q \psi^* (x, y, z)\, \psi (x, y, z)\, dx\, dy\, dz \qquad ...(2)$$

The probability that the electron is located somewhere must be unity.

$$\therefore \quad \iiint \psi^* \psi \, dx\, dy\, dz = 1 \qquad ...(3)$$

SCHRODINGER WAVE EQUATION

In 1926 Erwin Schrodinger developed a wave equation that describes the behaviour of atomic particles. Let us assume that a particle of mass 'm' is in motion along the x-direction. Let the wave function ψ be the dependent variable of the de Broglie wave which is a function of the coordinates x and t. Analogous to the classical wave, we may expect that, ψ will be a function of $(x - \omega t)$. As $\upsilon = \psi/k$, the wave function may be written as a function of $(kx - \psi t)$.

Using the relation $p = \hbar k$ and $E = \hbar \omega$, we can write

$$\psi = f\left(\frac{px - Et}{\hbar}\right) \qquad ...(1)$$

The more general wave would be a sum of a sine and cosine waves. Taking help of Euler's identity, we write the above equation in an exponential form as follows:

$$\psi = A \exp\left[\frac{i}{\hbar}(px - Et)\right] \qquad ...(2)$$

We assume that the energy and momentum of the particle are constant. Differentiating the above equation with respect to x, we get

$$\frac{\partial \psi}{\partial x} = A \frac{ip}{\hbar} \exp\left[\frac{i}{\hbar}(px - Et)\right]$$

$$\therefore \quad \frac{\partial \psi}{\partial x} = \frac{ip}{\hbar} \psi$$

Rearranging the terms in the above equation, we get

$$p\psi = \frac{\hbar}{i}\frac{\partial\psi}{\partial x} = -i\hbar\frac{\partial\psi}{dx} \qquad ...(3)$$

Differentiating the eq. (1) with respect to t gives

$$\frac{\partial\psi}{\partial t} = -\frac{iE}{\hbar} A \exp\left[\frac{i}{\hbar}(px - Et)\right] = -\frac{iE}{\hbar}\psi$$

Rearranging the terms in the above equation, we get

$$E\psi \frac{\hbar}{i}\frac{\partial\psi}{\partial t} = i\hbar\frac{\partial\psi}{\partial t} \qquad ...(4)$$

The partial derivatives with respect to x and t are connected by means of the relation between the energy and momentum. The classical expression for the kinetic energy in terms of the momentum is

$$E_k \frac{m\upsilon^2}{2} = \frac{(m\upsilon^2)}{2m} = \frac{p^2}{2m} \qquad ...(5)$$

The total energy and momentum are related by the expression

$$\frac{p^2}{2m} + V = E \qquad ...(6)$$

where Vis the potential energy of the particle. Multiplying the eqn. (6) with w, we obtain

$$\frac{p^2}{2m}\psi + V\psi = E\psi$$

Using the relations (41) and (42) into the above equation, we get

$$-\frac{\hbar^2}{2m}\frac{\partial^2\psi}{\partial x^2} + V\psi = i\hbar\frac{\partial\psi}{\partial t} \qquad ...(7)$$

The above equation is known as the *time-dependent Schrodinger wave equation.*

Knowing the form of V, eqn. (47) can be solved for the wave function ψ. In a number of cases the potential energy V of a particle does not depend on time; it varies with the position of the particle only and the field is said to be *stationary*. In the stationary problems Schrodinger equation can be simplified by separating out time and position- dependent parts. Accordingly, we can write the wave function as a product of x, ψ(x) and a function of t, φ (t).

We, therefore, write that

$$\psi(x, t) = \psi(x)\,\varphi(t) \qquad ...(8)$$

Equation (7) may be written as

$$-\frac{\hbar^2}{2m}\varphi\frac{\partial^2\psi}{\partial x^2}+V\psi\varphi=i\hbar\psi\frac{\partial\psi}{\partial t}$$

Dividing the above equation with ψ_φ, we get

$$-\frac{\hbar^2}{2m}\frac{1}{\psi}\frac{d^2\psi}{\partial x^2}+V=i\hbar\frac{1}{\varphi}\frac{d\varphi}{\partial t} \quad ...(9)$$

If we assume that the potential energy Vis a function of x only, the entire left hand side of eqn. (9) is a function of x only while the right hand side is a function of t only. Since x and t are independent variables, both the function of x and t must be equal to a constant. The constant that each side must equal is called the *separation constant* E. Thus,

$$-\frac{\hbar^2}{2m}\frac{1}{\psi}\frac{d^2\psi}{\partial x^2}+V=E \quad ...(10)$$

and $$i\hbar\frac{1}{\varphi}\frac{d\varphi}{\partial t}=E$$

Eq. (10) may be rewritten as

$$-\frac{\hbar^2}{2m}\frac{2^2\psi}{dx^2}+V\psi=E\psi$$

or $$\frac{2^2\psi}{dx^2}+\frac{2m}{\hbar^2}(V-E)\psi=0$$

$\therefore$ $$\frac{2^2\psi}{dx^2}+\frac{8\pi^2 m}{h^2}(V-E)\psi=0 \quad ...(11)$$

The above equation is called the time-independent Schrodinger equation.

Allowed Wave Functions and Energies

The time-independent wave equation is the pertinent equation for studying properties of atomic systems in stationary conditions. Wave mechanical methods of solving the problem of particle motion are essentially based on ψ functions. Appropriate wave equation is formulated by incorporating the particle mass 'w' and potential energy function V for the region in which the particle is located. The next step consists of solving the differential equation for solutions, namely for the ψ functions which will satisfy the differential equation. By solving the Schrodinger equation, we obtain the possible set of ψ functions. In case of bound particles the acceptable solutions for the differential equation are possible only for certain specified values of energy. These energy values will be the only possible results of precise measurements of the total energy of the particle. These discrete values of energy E_1, E_2, ... E_n are called eigen values or allowed values of the energy of the particle. The solutions ψ_1, ψ_2, ... ψ_n corresponding to the eigen energy values E_n are called the eigen functions. The quantization of energy thus appears as a natural element of the wave equation. Thus, using the Schrodinger equation, it is possible to determine first the electron energies and then the wave functions. These wave functions can then be used to determine the probability distribution function $\psi^*\psi$ of the electron, for various discrete energies as it revolves around the nucleus.

PROPERTIES OF SPECTRAL LINES

We record me spectral lines with the help of spectrometers, which have narrow slits. The spectral lines are therefore linear and possess three important properties; namely *position* measured in terms of frequency, finite *width*, which is a *range* of frequencies and *intensity*, the brightness of the line.

Width of Spectral Lines

Real atoms or molecules do not emit, radiation at precise frequencies; each emission is more or less broadened by various processes, and so each line is really a small package of slightly different frequencies. Therefore, even if we make the slit of the spectrometer infinitely narrow, there is nonetheless a minimum line width. The natural shape of an emission line appears under ideal conditions when the emitting atom is at rest and is not subjected to the external forces during the process of emission. In practice, several factors contribute to the line broadening. They are broadly classified into two categories:

(i) Homogeneous and

(ii) Inhomogenenous broadening. If the broadening mechanism affects each individual atom in the sample to the same extent, then the broadening is said to be *homogeneous*. In such a case all of the atoms in the sample will have the centre frequency v_o and the same line shape. Natural broadening and collision broadening belong to this category. On the other hand, if different atoms in the sample have slightly different frequencies for the same transition, the overall response of the sample broadens out and the broadening is said to be *inhomogeneous*. Doppler broadening and broadening due to crystal defects belong to this category.

1. *Collision broadening :* Atoms or molecules in liquid and gaseous phases are in continual motion and collide frequently with each other. When a radiation-emitting atom undergoes a collision, the emission conditions change and it is equivalent to an interruption. In the process, the phase of the emitted wave suffers random variations.
2. *Doppler broadening :* Doppler effect occurs when a source and an observer are in relative motion. The frequency as measured by the observer increases if the source and observer approach each other and decreases when they recede. In a gas due to the chaotic nature of thermal motion, all the directions of the molecules' velocities relative to a spectrograph are equally probable. Therefore, the radiation received by the spectrograph contains all the frequencies in the interval from v_0 $(1 - \upsilon/c)$ to v_0 $(1 + \upsilon/c)$, where v_0 is the frequency emitted by the molecules and υ is the velocity of thermal motion. Hence the spectral line is broadened. The broadening of the line caused by the Doppler effect is called *Doppler broadening.* In general, for liquids collision broadening is the most important factor, whereas for gases where collision broadening is less pronounced, the Doppler effect often determines the natural line width.
3. *Natural broadening :* Even in an isolated, stationary atom the energy levels are not indefinitely sharp. According to Heisenberg uncertainty principle, if a system exists in an energy state for a limited time Δt seconds, then the energy of that state will be uncertain to an extent ΔE where

$$\Delta E \times \Delta t \approx h/2\pi$$

Thus, we see that the ground state of a system is sharply defined since, left to itself, the system will remain in that state for an infinite time. Thus $\Delta t = \infty$ and $\Delta E = 0$. In contract, the lifetime of an excited state is about 10^{-8}s, which gives a value for ΔE of about 10^{-34}J.s$\div 10^{-8}$ s$=10^{-26}$ J. A transition between an excited state and the ground state will thus have an energy uncertainty of ΔE, and a corresponding uncertainty in the associated radiation frequency of $\Delta E/h$, which we can write as

$$\Delta v = \frac{\Delta E}{h} \approx \frac{1}{2\pi \Delta t} \quad ...(1)$$

If an excited electronic state lifetime is 10^{-8}s, then $\Delta v = 10^{8}$Hz. Thus, the natural broadening is relatively small in magnitude and it is often masked by other mechanisms.

The Intensity of Spectral Lines

The spectral line intensities are dependent on two factors: the *transition probability*, the likelihood of a system in one state changing to another state; and the *population of state*, the number of atoms or molecules initially in the state from which the transition occurs.

1. *Transition probability* : The detailed calculation of absolute transition probabilities involves a knowledge of the precise quantum mechanical wave functions of the two states between which the transition occurs. It is often possible to decide whether a particular transition is allowed or forbidden basing on selection rules.
2. *Population of states* : If we have two levels from which transitions to a third are equally probable, then obviously the most intense spectral line will arise from the level which initially has the greater population. At thermal equilibrium the population of a set of energy levels is governed by the Boltzmann law

$$\frac{N_2}{N_1} = e^{-\Delta E/kT} \qquad ...(2)$$

where N_1 is the population in the lower energy state E_1 and N_2 in the higher energy state E_2, $\Delta E = E_2 - E_1$, T is the temperature in K, and k is Boltzmann's constant.

WHAT IS LASER

Laser is one of the outstanding inventions of the 20th century. Laser is a *photonic device*, which is actually responsible for the resurgence of interest in optical technology and for the birth of a new field, namely *photonics*. The word 'LASER' is the acronym for Light Amplification through Stimulated Emission of Radiation. However, laser is *not* a simple amplifier of light but is actually a generator of light. In fact, the device should have been called a LOSER signifying Light Oscillation through Stimulated Emission of Radiation; this name was avoided because of its bad connotation and the name laser has been retained. Though it is a source of light, laser differs vastly from the traditional light sources. It is not used for illumination purposes as we use other light sources. Laser is more akin to radio and microwave transmitters and produces a highly directional coherent monochromatic light beam. Laser is the most sought-after tool in metal-working, entertainment electronics, optical communications, bloodless surgery, weapon guidance in wars and in a wide variety of other fields.

SOME THEOREMS IN PHOTOMETRY

In simple situations several inter-relations hold good between the quantities defined above. We will see some of them which are used widely.

(a) Relation $E = I \cos \theta/r^2$. If the dimensions of the source are much smaller than the distance of the observer, then the source is called a 'point source.' In such cases, the solid angle $\Delta\Omega$ subtended by a given surface of area ΔS at the source is defineable: We have $\Delta\Omega = \Delta S \cos \theta/r^2$, where θ is the angle between the ray and the normal to the surface (Fig. 6.1). If the source has intensity led in the concerned direction, the flux received is $Id\Omega$ lm, and illumination is

$$E = \frac{Id\Omega}{dS} = \frac{IdS \cos \theta}{r^2 dS} = \frac{I \cos \theta}{r^2} \quad ...(1)$$

This is one of the Lambert relations. It combines the inverse square law with $\cos \theta$ dependence. *But it is useful only if the source can be treated as a point source.*

(b) Relation $L = \pi B$. Fora surface of uniform brightness B, the luminance L is given by πB. The elementary solid angle for the hollow cone between θ and $\theta + d\theta$ from the normal to a surface is $2\pi r \sin \theta.\ rd\theta \div r^2 = 2\pi \sin \theta\ d\theta$. The brightness is defined by

$$B = \frac{\Delta I}{\Delta S \cos \theta} \quad ...(2)$$

Hence the luminous flux in the hollow cone due to area ΔS is $B\ \Delta S \cos \theta.\ 2\theta \sin \theta\ d\theta$. Integrating this over the whole available solid angle gives $L.\Delta S$.

Thus

$$L\Delta S = \int_0^{\pi/2} B\ \Delta S \cos \theta\ 2\pi \sin \theta\ d\theta$$

If B is taken as independent of θ, we get the result

$$L = \pi B \quad ...(3)$$

This is another *Lambert relation.* One consequence of this is that luminosity of the blackbody surface used for defining the Candela is $6\pi \times 10^5$ *l*m/m^2 [check it].

(c) *Brightness of a surface is independent of the distance of the viewer*. This ts because the illuminance falls as $1/r^2$, and the solid angle subtended at the eye by a given area of the surface also falls as $1/r^2$. Hence the (lux received *per unit solid angle* remains unchanged. Of course, this assumes that there is no attenuation in the medium.

(d) Relation $E = \pi L$: The illuminance at the centre of a flat surface under a hemispherical cover" of uniform luminance L is given by πL. An elementary ring on the hemisphere defined by θ and $\theta + d\theta$ has area $2\pi r \sin\theta \, r d\theta$ and hence intensity L. $2\pi r^2 \sin\theta \, d\theta$. Its contribution to the illuminance at the centre is $\cos\theta/r^2$ times this (Eqn. 6.1)

That gives

$$dE = L.2\pi \sin\theta \cos\theta \, d\theta$$

Integration from $\theta = 0$ to $\theta = \pi/2$ gives ...(4)

$$E = \theta L$$

The result is independent of r.

The illuminance over the surface due to a parallel infinite surface of luminance L is also πL. The proof of this we will leave for the reader. The result is independent of the distance between the two surfaces.

(e) *Relation* I = BA For a surface of uniform brightness B, the intensity for a distant observer is B.A, where A is the area of the source projected on a plane transverse to the line of sight. This follows directly from Eqn. (2).

$$dI = BdS \cos\theta = BdS'$$

$$I = B\int dS' = B.A. \qquad ...(5)$$

The relation has special interest in the viewing of stars.

(f) *No image can have brightness greater than that of the source.* the geometry in image formation, with areas S, S' and solid angles Ω, Ω' shown, besides u and v.

Then we have from geometry

$$S' = \frac{\upsilon^2}{u^2} S; \; \Omega' = \frac{u^2}{\upsilon^2} \Omega$$

If we consider a source of uniform brightness B, then the luminous flux collected by the lens is B.S.Ω and this spreads over area S', so that *illuminance* E on the image plane is

$$E = B\,\Omega\,\frac{S}{S'} = B\frac{A}{\upsilon^2} = B\,\frac{A}{\upsilon^2} \qquad ...(6)$$

where A is the area of the lens. Thus E can be increased by having larger lens area and smaller focal length (to give lower υ).

But consider *brightness* B' of the image. The flux collected by the lens emerges in a solid angle Ω'. Hence the flux per unit area of the image per unit solid angle (which is B') comes to

$$B' = \frac{BS\Omega}{S'\Omega'} = B\left(\frac{S}{S'}\right)\left(\frac{\Omega}{\Omega'}\right) = B\,\frac{u^2}{\upsilon^2}.\frac{\upsilon^2}{u^2} = B \qquad ...(7)$$

Thus the best we can get is B' = B; in practice losses due to various effects give B' < B.

SOME TYPICAL INTENSITY AND ILLUMINANCE VALUES

For comfortable work in different situations the illuminations recommended are as follows:

Home, libraries, office, etc.	:	100-300 lx
Sewing and finer work	:	500-1000 lx
Operation theatre table	:	5000-8000 lx

The accommodation provided by nature in the eye is able to take care if there is excess illumination; but in poor illumination one has less than the optimum efficiency of work.

A typical 100 W bulb of incandescent filament type has intensity ~200 cd in the forward direction, so that the illuminance 2m below it would be ~50 lx due to direct light Multiple scattering from walls of the room raise this to 2 to 5-fold, depending on the wall colour and room size. Sunlight at mid-day produces an illuminance about 10^5 lx. In fullmoon at mid-night it is about 0.25 lx. That we can at least see, if not read, under such a large range of illuminations is remarkable.

A sodium vapour lamp has much larger intensity than an incandescent filament lamp; but the filament could be much brighter than the sodium vapour lamp, because the area is very much smaller in the second case. Any illuminated surface acts as a secondary source of luminous flux, defined by brightness B and luminance L, just as for a primary source. Typically, a matt surface coated with MgO will have luminance L about 95% of the illuminance—and also have B equal in all directions. A mirror reflector, on the other hand, will have B. zero in all directions except the range of specular reflection and L around 80-98% of E, depending on the quality of the reflecting polish.

LUMINOSITY AND LUMINOUS EFFICIENCY OF A SOURCE

A primary source of light consumes energy at the rate of X wait. Out of this a certain part, X' watt, goes out as radiant energy, the rest is consumed otherwise. This X' watt creates a total luminous flux ϕ lm. Then the ratio (ϕ/X') lm/W is called the luminosity of the source, and the ratio (ϕ/X) lm/W is called the *luminous efficiency* of the source. Not unoften, the distinction between X and X' is omitted and one writes

$$\eta = \frac{\text{Lu min ous flux (lm)}}{\text{Power concerned (W)}} \quad ...(1)$$

where η is used for both luminosity and luminous efficiency.

The value of η defined as ϕ/X', depends on the spectral distribution of the radiation. Obviously, if all the radiation falls outside the visible range (~4000 to ~7000^1), then $\phi = 0$ and hence $\eta = 0$. Within the visible range, ϕ/X' is wavelength-dependent, being the maximum for the green region λ~5550 A°, and falling off with λ both ways. This maximum value is

$$\eta_{max} = 683 \text{ lmlW} \qquad \text{at } \lambda = 5550\text{A}$$

The number 683 comes from the way 1 candela is defined. The relative spectral sensitivity, K_λ of the human eye.

It will be seen that if a source emits all its radiation at λ = 5550Å, it will have η = 683 lm/W. If the spectral distribution (intensitywise) matches the curve of Fig. 6.3, then η = 683 × averaged K_λ, + 200 lm/W. For a general case it would be

$$\eta = \frac{\int K_\lambda P_\lambda d_\lambda}{\int P_\lambda d_\lambda} \times 683 \text{ lm/W} \quad ...(2)$$

where P_λ is the radiated power at λ. This is for η defined by ϕlX'. The realistic η is defined by ϕ/X and it will be lower, depending upon the power (X–X') consumed otherwise by the concerned system.

Sunlight happens to have its radiant energy maximum close to 5550A°, and it appears that nature has adapted the human eye to match its sensitivity curve to secure the best from sunlight. For other sources the typical η values are as follows:

Incandescent lamp	:	14 lm/W
Fluorescent lamp	:	43 lm/W
Halogen lamp	:	58 lm/W

The values mean that a 100 TV incandescent lamp,would emit ~1400 lm flux. Considered isotropic, its intensity would be 1400/4π 120 cd. With a 'shade,' the I value in the front direction may reach ~400 cd; from a 2m distance it would create illuminance of ~ 100 lx.

PHOTOMETERS

In general a photometer compares the luminous intensities of two sources by matching the brightness produced by them on a surface. The steps involved are (a) illumination produced = $I \cos \frac{\theta}{r^2}$; (b) luminance-$= f.I \cos \frac{\theta}{r^2}$; (c) brightness = ff' I cos θ/(r^2 cos θ'). Here f and f' are appropriate fractions, and θ and θ' are angles of incidence and viewing. If θ is kept zero and f,f', θ' are identical for the two cases, then equality of brightness means

$$\frac{I_1}{r_1^2} = \frac{I_2}{r_2^2} \Rightarrow \frac{I_1}{I_2} = \left(\frac{r_1}{r_2}\right)^2 \qquad ...(1)$$

An assumption involved is that the concerned sources are 'point sources,' which means that r values are much greater than the size of either source.

Different photometers use different methods of matching the brightness. We describe some of them.

(a) *Bunsen 's Grease-Spot Photometer :* It uses a mat-white paper (like a filter paper) with a central greased spot. The ungreased part gives diffuse scattering of light falling on the same face, while the greased part gives diffuse scattering of light falling on the opposite face. If the distances of the sources are adjusted to make the grease spot vanish in the background of the ungreased part, Eqn. (9a) gives I_1/I_2.

If we assume that the ungreased part scatters a fraction a, and the greased part scatters fraction b and c of the light falling on the same and the opposite face, respectively, then the brightness of face 1 is

$$\left(\frac{I_1}{r_1^2}\right)a \qquad \text{for the ungreased part}$$

$$\left(\frac{I_1}{r_1^2}\right)b + \left(\frac{I_2}{r_2^2}\right)c \quad \text{for the greased part} \qquad ...(2)$$

Equality of brightness will still lead to Eqn. 9a if b + c = a (*i.e.*, there is no loss by absorption). In practice, the grease spot never fully vanishes in the background, and what one has to adjust is the 'identical view' of the two faces of the detector paper.

(b) *Lummer-Brodhun Photometer :* In this also the two sides of a mat white surface (a thick one, this time) are illuminated by light from the two sources under study, but comparison of brightness of the two faces is made with the help of a 'photometer cube' C. In this 'cube' an air space separates the hypotenuse faces of two prisms, but a central portion is sealed with Canada balsam, which is transparent, like glass. Thus an observer views face 1 through the central portion of C and face 2 through the outer portion of C. In effect, this is like setting b = 0 and c = a in relations 10. Because total internal reflection gives 100% reflection, these conditions hold well, assuming that the faces of all the right-angled prisms are kept very clean.

(c) *Flicker Photometers :* If the sources under comparison give out light of different colours, then the aforesaid methods fail. A flicker photometer is then used. In principle, the two surfaces, illuminated separately by the two sources, are kept in the same line of sight, but an arrangement is made that these two are seen alternately at a fairly large frequency. The result is that what one sees is neither colour of source 1 nor that of source 2, but a mix. What is important is that if the visual sensations of the two are equal, then no flicker is observed. A crucial fact is that at low frequency of the alternating view, the colon r mixing (in the sensation at the eye) does not take place, and at very high frequency of alternating view the flicker is not observable even when brightnesses are unequal (due to persistence of vision). So a careful balance is needed. One starts from low rate and goes on to higher ones to see if the colour distinction becomes absent; and then one adjusts the distances till the flicker is eliminated.

Schematic of *Guild's flicker photometer.* Surfaced is illuminated with source 1 and seen at 45° from its normal. Surface B is a sector disc illuminated with source 2 and seen again at 45° from the normal. As the sector disc is rotated, the eye alternately views A and B. If both the surfaces have identical diffuse coating (say, of MgO), which is neutral for colours, then absence of flicker means $\frac{I_1}{r_1^2} = \frac{I_2}{r_2^2}$.

The sectoral disc B is rotated so that alternately surfaces A and B are seen by the eye. The viewing chamber has a diaphragm D painted white and kept illuminated to serve as a comfortable background against which A/B is seen ($r_1 = a + b$).

(d) *Ulbright's Integral Photometer :* The luminous intensity of a source is often direction-dependent. One may therefore measure I in different orientations and then find the averaged I over the total solid angle. But Ulbright's photometer does the integration in another way. Source S placed inside a hollow sphere, whose inside walls are perfectly diffusing, mat painted typically with MgO. Due to multiple reflections within, the radiation inside becomes isotropic. If a window W is placed on the wall, the output flux there is equal to the total flux emitted by the source. Comparing the brightness of an opal glass placed at this window with that produced by a standard source from a measured distance gives the desired result.

OSCILLATING ELECTRIC DIPOLE

Let us consider an oscillating dipole which consists of equal positive and negative charges +q and –q located at A and B respectively. The charges vibrate along the line AB with simple harmonic motion, the charge of one sign being out of phase with that of the other by 180°. Let us further assume that the oscillations occur at constant amplitude and there is no damping.

If A_1 is the amplitude of the vibration of the charge + q, the displacement of the charge at any instant is

$$z_1 = A_1 \cos \omega t$$

The displacement of the negative charge vibrating with the same frequency but out of phase by 180° with the first is given by

$$z_2 = A_2 \cos (\omega t + \pi) = -A_2 \cos \omega t$$

where A_2 is the amplitude of vibration of the second charge. At any instant the distance between the charges is

$$z_2 - z_2 = (A_1 + A_2) \cos \omega t$$

The electric moment of the dipole is

$$q (z_1 - z_2) = p_o \cos \omega t \text{ where } p_o = q (A_1 + A_2)$$

According to Maxwell's equations, the electric intensity at any point around the dipole is made up of two components. Close to the dipole the field at any instant may be treated as if static. At a distance r from the centre of the dipole large compared with AB

$$E_r = \frac{2 p_o \cos \theta}{r^3} \cos \omega t$$

and

$$E_\theta = \frac{p_o \sin \theta}{r^3} \cos \omega t$$

The magnetic field near the dipole is

$$H = \frac{\omega p_o \sin \theta}{cr^2} \sin \omega t$$

at right angles to the rz-plane. The lines of magnetic force are therefore circles about the z-axis.

At a great distance from the dipole E_θ. has a negligible magnitude compared with –g, and the electric field intensity is given by

$$E_\theta = \frac{4\pi^2 p_o \sin \theta}{\lambda^2 r} \cos (\omega t - kr) \qquad ...(1)$$

and the magnetic intensity is given by

$$H_\Phi = \frac{4\pi^2 p_o \sin \theta}{\lambda^2 r} \cos (\omega t - kr) \qquad ...(2)$$

in a direction at right angles to the rz-plane, *i.e.*, in the direction of increasing azimuth Φ measured about the z-axis. Therefore, the lines of magnetic force are circles around the z-axis. Both E_θ and H_Φ are zero along the axis of the dipole on account of the factor sin θ and are maximum in the equatorial plane. The radiation is emitted from the dipole in greatest intensity in directions at right angles to the line of oscillation.

The Pointing flux is

$$S = \frac{c}{4\pi} E_\theta H_\Phi \frac{4\pi^2 c p_o^2 \sin^2 \theta}{\lambda^2 r^2} \cos^2 (\omega t - kr) \qquad ...(3)$$

in the direction of radius vector.

The energy radiated per unit time is given by

$$I = \int_0^{\pi} S.2\pi r^2 \sin\theta \, d\theta = \frac{32\pi r^4 \, cp_o^2}{3\lambda^2} \cos^2 (\omega t - kr)$$

The mean rate of radiation is

$$I_{ave} = \frac{16\pi^4 c \, p_o^2}{3\lambda^4} = \frac{16\pi^4 \, v^4 \, p_o^2}{3c^3} \quad ...(4)$$

QUANTUM OPTICS

Einstein postulated that electromagnetic radiation is made up of photons indicating thereby that the electromagnetic field itself is quantized. Photons are the basic and discrete units of energy. Each photon has an energy E = hv and momentum $p = \frac{hv}{c} = \left(\frac{h}{2\pi}\right)k$. Photons are stable, electrically neutral and massless elementary particles. They always travel with the speed of light 'c' and exist only at that speed.

A photon has the mass $m = \frac{hv}{c^2}$ which is the mass of the electromagnetic field and is not associated with the rest mass because photons at rest do not exist. Photons always travel with the speed of light whether in vacuum or in a medium. Note that the velocity of photons in a medium is different from the velocity of the propagation of the wave front of light in the medium. Photons are emitted when atoms make a transit from a higher energy state to a lower energy state. They are also emitted upon the acceleration or retardation of charged particles, in the decay of small particles, and in the annihilation process of electron-positron pairs.

Photons have spin angular momentum of 1η and obey Bose-Einstein statistics. Therefore, they are called *bosons*. Bosons do not obey Rauli exclusion principle and therefore any number of photons can occupy the same state. When a very large number of photons occupy the same state, the inherent discreteness of the light beam disappears and the beam appears to be a bunch of continuous electromagnetic waves. A monochromatic plane wave may be regarded as a stream of photons having high population density, all photons occupying the same state.

There is an intimate relationship between the wave and corpuscular nature of photons. Photons exhibit interference and diffraction phenomenon which are the result of their redistribution in space. The intensity of light reaching any point of space is a measure of the number of photons striking this point. The quantum properties of light are due to the fact that the energy, momentum and mass of electromagnetic radiation is concentrated in photons.

Quantum optics deals with the discrete nature of the emission, propagation and interaction of light with a substance. It is chiefly concerned with the generation of coherent light and detection of light. Lasers, light emitting diodes (LEDs) and various types of photo-detectors are quantum optic devices.

THERMAL RADIATION

In majority of the cases, light is produced as a consequence of their temperature. Therefore, they are said to be thermal sources and the radiation emitted by the source by virtue of its temperature is called *thermal radiation*. Thermal radiation is electromagnetic in nature and its energy is smoothly distributed over all wavelengths. Therefore, a thermal source produces *continuous spectrum*.

The intensity and the predominant wavelength of radiation vary with the temperature of the body. At low temperatures the radiation mainly lies in the infrared region. As the temperature of the body is increased, the component of maximum intensity shifts to a higher and higher frequency. For example, the filament of an incandescent bulb appears dark at room temperature and as current is passed, it gets heated up. As the current increases through the filament, it appears initially red, orange gradually and then yellow and finally it emits white light. At temperatures above 1000°C, a heated body is capable of giving out energy in the form of waves of all possible wavelengths. Normally the amount of radiation emitted by a hot body depends on factors such as the properties of its surface.

Kirchhoff's Law

Bodies radiate energy either in the invisible form or in the visible form. Similarly, they absorb light to varying degrees. A body that is a good light-absorber appears black. In 1859, Gustav Kirchhoff formulated the following law of radiation.

If a body is in thermal equilibrium, the energy it absorbs equals the energy it gives away in the form of radiation.

The law is expressed mathematically as

$$\frac{E(v,T)}{A(v,T)} = \varepsilon(v,T) \qquad ...(1)$$

where E (v, T) is the emissive power of the body, A (v, T) is its absorptive power and ε (v, T) is the emissive power of a perfect blackbody.

A *perfect blackbody* is a body that absorbs all the radiation that is incident on it, no matter what the radiation frequency is. Conversely, when a perfect blackbody is heated it emits radiation at all frequencies. Thus, it is a good radiator as well as a good absorber.

Transparent and reflective materials are poor radiators and poor absorbers. *An object that is a good radiator at a given wavelength is also a good absorber at the same wavelength.* This is another form of Kirchhoff's law.

The emissive power represents the ability of a body to radiate efficiently and is also associated with its ability to absorb radiation. Only for a true blackbody, $\varepsilon = 1$. For other bodies, the emissivity usually lies between 0.2 and 0.9, and for highly polished metals e may be as low as 0.01.

Laws of Blackbody Radiation

(i) Stefan-Boltzmann Law

The Stefan-Boltzmann law is an empirical relationship obtained by Stefan and later derived theoretically by Boltzmann. It states that the total radiation emitted from a blackbody at temperature T is proportional to the fourth power of the absolute temperature of the body T^4.

$$I = \sigma T^4 \qquad ...(2)$$

where σ is called Stefan's constant having a numerical value of 5.67×10^{-8} W/m^2–K4.

(ii) Wien's Law

If we plot the distribution of radiant energy as a function of wavelength at different temperatures, we obtain a set of curves as shown in Fig. 21.2. All curves have a peak and the peak is displaced towards the shorter wavelengths as the temperature rises. Wien's displacement law states that the peak wavelength, λ_{max} at which the maximum emission occurs for any given temperature is inversely proportional to the absolute temperature of the body. Thus,

$$\lambda_{max} = \frac{2.8978 \times 10^{-3}}{T} \text{ mK} \quad ...(3)$$

The shift to shorter wavelengths agrees with common experience.

ATTENUATION OF LIGHT IN AN OPTICAL MEDIUM

When light travels through a medium, a gradual reduction in its intensity occurs mainly because of the processes of absorption and scattering of light in the medium.

(i) Light *absorption* occurs because part of the incident light is transformed into the energy of motion of the atoms in the medium; and

(ii) light is scattered when it encounters obstacles of sizes smaller than a wavelength. The reduction in intensity with distance in a medium is called *attenuation of light*. The following relation governs the attenuation of 14 a transparent medium.

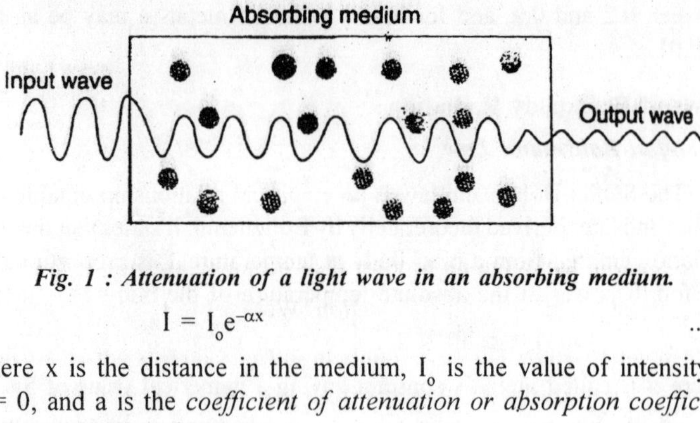

Fig. 1 : Attenuation of a light wave in an absorbing medium.

$$I = I_0 e^{-\alpha x} \quad ...(1)$$

where x is the distance in the medium, I_0 is the value of intensity at x = 0, and a is the *coefficient of attenuation or absorption coefficient* of the material at frequency v.

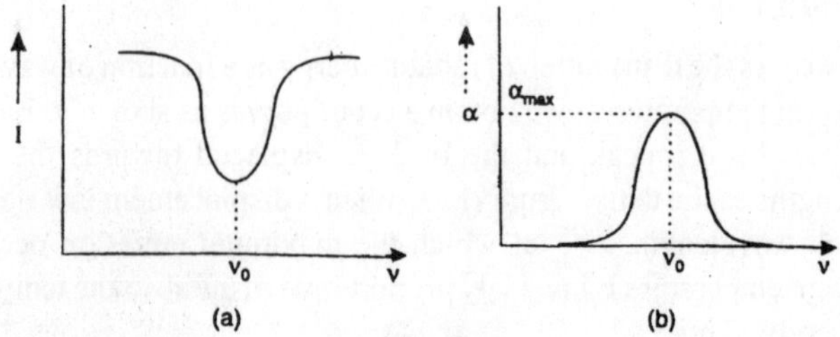

Fig. 2

Fig. 2(a) shows the variation of intensity I with frequency V at a fixed distance x in the medium. Fig. 2(b) shows the variation of the attenuation coefficient a with frequency v. In general a is a positive quantity and we say that the material has & positive *coefficient of absorption.*

INTERACTION OF LIGHT WITH MATTER

The process of the transfer of energy from atom to light is not possible from classical point of view. However, the possibility arises if the interaction of light with medium is considered from the point of view of quantum mechanics. A *laser* is a light source that utilizes the quantum processes for its operation. It is therefore, necessary to appreciate the quantum processes involved in the development of a laser. The radiation incident on a material is viewed as a stream of photons, where each photon carries an energy E = hv. We assume that the two energy levels of the atoms in the material have an energy difference $(E_2 - E_1) = hv$. When photons travel through the medium, three different processes are likely to occur. They are absorption, spontaneous emission and stimulated emission. We study these in detail.

Absorption

Suppose an atom is in the lower energy level E_1. If a photon of energy $hv = (E_2 - E_1)$ is incident on the atom, it imparts its energy to the atom and disappears. Then we say that the atom absorbed an incident photon. As a result of absorption of adequate energy, the atom jumps to the excited state E_2. The transition is called an *absorption transition.* It is also referred to as *induced absorption.* We may express the process as

$$A + hv = A^*$$

where A is an atom in the lower state and A* is an excited atom.

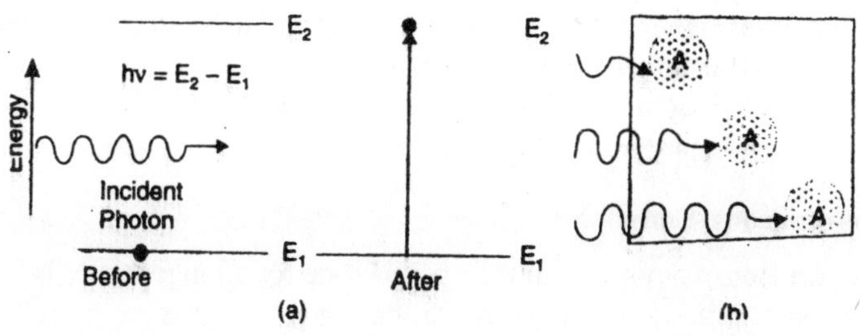

Fig. 3 : Absorption process (a) Induced absorption (b) Material absorbs photons.

In each absorption transition event, an atom in the medium is excited and one photon is subtracted from the incident light beam, which results in attenuation of light in the medium.

The number of atoms per unit volume that undergo absorption transitions per second is called the *rate of absorption transition.* It is denoted by

$$R_{abs} = -\frac{dN_1}{dt} \qquad ...(1)$$

where $\left(-\frac{dN_1}{dt}\right)$ represents the rate of decrease of population at the lower level E_1. The rate of absorption can also be represented by the rate of increase of population at the upper level E_2, as

$$R_{abs} = \frac{dN_2}{dt} \qquad ...(2)$$

The number of absorption transitions occurring in the material at any instant are proportional to the number of atoms in the lower level E_1 and the density of photons in the incident beam. When the atoms are more at the lower energy level, then more atoms can jump into the excited state.

Similarly, when more photons are incident on the assembly of atoms, then more atoms can get excited to the higher energy level. Then the rate of absorption transition is given by

$$R_{abs} = B_{12}\rho(v)N_1 \qquad ...(3)$$

where N_1 is the population of atoms at E_1, $\rho(v)$ the energy density of the incident beam and B_{12} is the constant of proportionality. B_{12} is known as the Einstein coefficient/or induced absorption. It indicates the probability of occurrence of an induced transition from level $1 \rightarrow 2$.

Induced absorption involves the excitation of atom to the fixed higher level only. As a result of this absorption N_1 decreases and N_2 increases. But under normal conditions N_2 cannot be greater than N_1. Therefore, as light propagates through the medium, it gets absorbed. However, N_2 can be made greater than N_1 using special techniques.

Spontaneous Emission

An atom cannot stay in the excited state for a longer time. In a time of about 10^{-8}s, the atom reverts to the lower energy state by releasing a photon of energy hv, where $hv = (E_2 - E_1)$. The emission of photon occurs on its own and without any external impetus given to the excited atom (Fig. 4). Emission of a photon by an atom *without any external impetus* is called *spontaneous emission.* We may write the process as

$$A^* \rightarrow A + h\nu$$

The number of spontaneous transitions depends only on the number of atoms N_2 at the excited state E_2. Therefore, the rate of spontaneous transitions is given by

$$R_{sp} = A_{21} N_2 \quad ...(4)$$

where A_{21} is the proportionality constant and is called the *Einstein coefficient for spontaneous emission.* A_{21} represents the probability of a spontaneous transition from level 2 → 1. It is to be noted that the process of spontaneous emission is independent of the incident light energy.

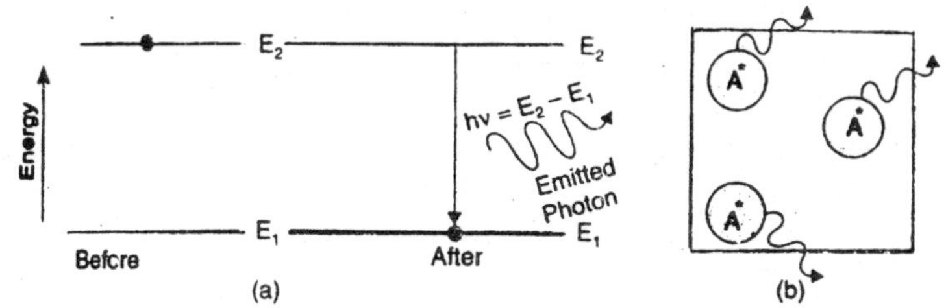

Fig. 4 : Spontaneous emission (a) emission process (b) Material emits photons haphazardly.

It follows from quantum mechanical considerations that spontaneous transition takes place from a given state to states lying lower in energy. Thus, spontaneous transition is not possible from level E_1 to level E_2. Therefore, the probability of spontaneous transition from E_1 to E_2 is zero.

$$\therefore \quad A_{12} = 0 \quad ...(5)$$

Let now look at the salient features of spontaneous emission of light.

Characteristics of spontaneous emission

(i) The process of spontaneous emission is essentially *probabilistic* in nature and is not amenable for control from outside.

(ii) The instant of transition, direction of propagation, the initial phase and the plane of polarisation of each photon are all *random.*

(iii) The light resulting through this process is *not monochromatic.*

(iv) As different atoms in the source emit photons in different directions, light spreads in all directions around the source. The light intensity goes on decreasing rapidly with distance from the source.

(v) Light emitted through this process is *incoherent*, as it results from a superposition of wave trains of random phases. The net intensity is proportional to the number of radiating atoms. Thus,

$$I_{total} = NI \qquad ...(6)$$

where N is the number of atoms and I is the intensity of light emitted by one atom. It is the process of spontaneous emission that dominates in conventional light sources.

Einstein's Prediction

It is seen that the rate of spontaneous transitions is determined only by the population N_2 at the higher energy level whereas the rate of absorption transitions is determined by the population N_1 at the lower energy level and the energy density $\rho(v)$ in the incident light. If absorption and spontaneous emission were the only processes operative, then obviously the number of atoms absorbing radiation per second would be more than the number of atoms emitting light per second. Eventually, we may end up with a non-equilibrium situation where all the atoms in the medium are excited. But this condition is not observed in practice. It means that equilibrium is maintained. Therefore, in order to account for the state of equilibrium between light and matter, Einstein pointed *out that if a photon can stimulate an atom to move from a lower energy level E_1 to a higher energy level E_2 by means of absorption transition, then a photon should also be able to stimulate an atom from the same upper level E_2 to the lower level E_1*. This alternative mechanism of photon emission depends on the photon density present in the medium and is known as *stimulated emission*.

Stimulated Emission

An atom in the excited state need not "wait" for spontaneous emission of photon. Well before the atom can make a spontaneous transition, it may interact with a photon with energy $hv = E_2 - E_1$ and make a downward transition. The photon is said to stimulate or induce the excited atom to emit a photon of energy $hv = (E_2 - E_1)$. The passing photon does not disappear and in addition to it there is a second photon which is emitted by the excited atom. The phenomenon of forced photon emission by an excited atom *due to the action of an external agency* is called *stimulated emission or induced emission.* The process may be expressed as

$$A^* + h\nu \rightarrow A + 2h\nu$$

The rate of stimulated emission of photons is given by

$$R_{st} = B_{21}\,\rho(\nu)N_2 \qquad ...(7)$$

where B_{21} is the *Einstein coefficient for stimulated emission* and represents the probability for induced transition from level 2 → 1.

Characteristics of Stimulated emission

(i) The process of stimulated emission is *controllable* from outside.

(ii) The photon induced in this process propagates in the *same direction* as that of stimulating photon.

(iii) The induced photon has features identical to that of the inducing photon. It has the *same frequency, phase and plane of polarisation* as that of the stimulating photon.

(iv) *Multiplication of Photons :* The outstanding feature of this process is the *multiplication of photons*. For one photon interacting with an excited atom, there are two photons emerging. The two photons travelling in the same direction interact with two more excited atoms and generate two more photons and produce a total of four photons. These four photons in turn stimulate four excited atoms and generate eight photons, and so on. The number of photons builds up in an avalanche like manner.

(v) Light amplification: All the light waves generated in the medium are due to one initial wave and all of the waves are in phase. Thus, the waves are coherent and interfere constructively.

The net intensity of light will be proportional to the square of the number of atoms radiating light. Thus,

$$I_{total} = N^2 I \qquad ...(8)$$

The light emitted through the process of stimulated emission is of very high intensity and we say light is *amplified.* At any reasonable temperature there cannot be enough atoms in excited states for any appreciable amount of stimulated emission from these states to occur. Rather absorption is much more probable.

SEMICONDUCTOR LASER

A *semiconductor diode laser* is a specially fabricated pn junction device, which emits coherent light when it is forward biased. R.N. Hall and his coworkers made the first semiconductor laser in 1962. It is made from Gallium arsenide (GaAs) which operated at low temperatures and emitted light in the near IR region. Semiconductor lasers working at room temperature and in continuous wave mode are produced by 1970. Now pn-junction lasers are made to emit light almost anywhere in the spectrum from UV to IR. Diode lasers are remarkably small in size (0.1mm long). They have high efficiency of the order of 40%. Modulating the biasing current easily modulates the laser output. They operate at low powers. In spite of their small size and low power requirement, they produce power outputs equivalent to that of He-Ne lasers. The chief advantage of a diode laser is that it is portable. Because of the rapid advances in semiconductor technology, diode lasers are mass produced for use in optical fibre communications, in CD players, CD-ROM drives, optical reading, high speed laser printing etc wide variety of applications.

A *semiconductor* is a material with electrical properties intermediate to those of a conductor and an insulator. The allowed energy values of the valence electrons in semiconductors occur within two well-defined energy bands separated by an energy gap known as *band gap*. A pure semiconductor crystal has exactly enough electrons to fill all the states in the lower band, namely valence band. However, when a covalent bond is just broken, an electron is just set free. Then we say that the electron jumped into the upper band, namely *conduction band.*

The electron jumping to the conduction band leaves behind a vacancy in the valence band. The vacancy is called a hole and is assigned a positive charge and a mass equivalent to that of an electron. In a pure semiconductor, for each covalent bond broken an electron and a hole are generated. Therefore, the number of electrons in the conduction band and the number of holes in the valence band are equal. When a conduction electron falls into the valence band, it recombines with a hole there. The electron rejoins the broken covalent bond and therefore both the electron and hole disappear. The recombination energy is released in the form of heat in silicon and germanium crystals. In some crystals it is released in the form of light.

Doping with small amounts of impurities can drastically increase the electrical conductivity of a pure semiconductor. When the dopant is a pentavalent element, each dopant atom contributes an electron to the conduction band without creating a hole simultaneously in the valence band. Hence the addition of the pentavalent element increases the number of conduction electrons which become the *majority carriers* in the silicon crystal. As negatively charged electrons are current carriers in this crystal, it is called a *n-type semiconductor*. On the other hand, a trivalent dopant atom produces a hole in the valence band without the simultaneous generation of electron in the conduction band. Hence the addition of the trivalent element increases the number of holes which become the *majority carriers* in the silicon crystal. As positively charged holes are current carriers in this crystal, it is called a *p-type semiconductor*.

There is a reference level in the energy band diagram of each type of semiconductor. The reference level is called the *Fermi level*. The Fermi level Ep is nearer to the top of the valence band in the p-type material and the Fermi level E_{Fn} is nearer to the bottom of the conduction band in the n-type material. When the p-type and n-type materials are joined at the atomic level to form a pn-junction device, equilibrium is attained only when equalization of Fermi levels takes place. The energy levels in p-region move up and those in n-region move down till the Fermi levels (E_{Fn} and E_{Fn}) in both the regions come to the same level. The mutual displacement of the energy levels on both sides of the junction causes a bending of the energy bands around the junction.

Achieving Population Inversion in a Semiconductor

Population inversion is required for producing stimulated emission. The way in which population inversion is achieved in semiconductors is very different from the way it is established in other types of lasers. A semiconductor cannot be regarded as two-level atomic system. It consists of electrons and holes distributed in the respective energy bands. Therefore, laser action in semiconductors involves energy bands rather than discrete levels. Secondly, in other types of lasers, population inversion is obtained by exciting electrons in spatially isolated atoms. In semiconductors, electrons are not associated with specific atoms but are injected into the conduction band from the external circuit. Therefore, the conduction band plays the role of excited level while the valence band plays the role of ground level. Population inversion requires the presence of a large concentration of electrons in the conduction band and a large concentration of holes in the valence band. A simple way to achieve population inversion is to use a semiconductor in the form of a pn-junction diode formed from heavily doped p- and n-type semiconductors.

PN-Junction Laser

Construction : Fig. 5 shows the schematic of a semiconductor laser. A simple diode makes use of the same semiconductor material, say, GaAs on both sides of the junction. Starting with a heavily doped n-type GaAs material, a p-region is formed on its top by diffusing zinc atoms into it. A heavily zinc doped layer constitutes the heavily doped p-region. The diode is extremely small in size. Typical diode chips are 500 μm long and about 100 μm wide and thick. The top and bottom faces are metallized and metal contacts are provided to pass current through the diode. The front and rear faces are polished parallel to each other and perpendicular to the plane of the junction. The polished faces constitute the Fabry-Perot resonator. In practice there is no necessity to polish the faces. A pair of parallel planes cleaved at the two ends of the pn-junction provides the required reflection to form the cavity. The two remaining sides of the diode are roughened to eliminate lasing action in that direction. The entire structure is packaged in small case which looks like the metal case used for discrete transistors.

Working : The energy band diagram of a heavily doped pn-junction. Because of very high doping on n-side, the donor levels are broadened and extend into the conduction band. The

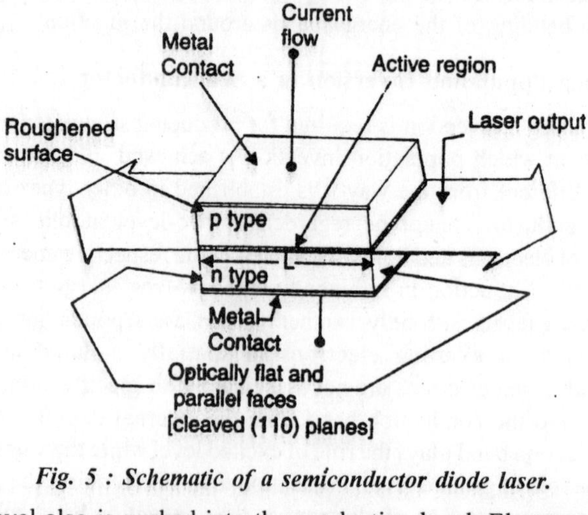

Fig. 5 : Schematic of a semiconductor diode laser.

Fermi level also is pushed into the conduction band. Electrons occupy the portion of the conduction band lying below the Fermi level. Similarly, on the heavily doped p-side the Fermi level lies within the valence band

and holes occupy the portion of the valence band that lies above the Fermi level. At thermal equilibrium, the Fermi level is uniform across the junction.

When the junction is forward-biased, electrons and holes are injected into the junction region in high concentrations. In other words, carriers are *pumped* by the dc voltage source. At low forward current level, the electron-hole recombination causes spontaneous emission of photons and the junction acts as an LED. As the forward current through the junction is increased the intensity of the light increases linearly. However, when the current reaches a threshold value the carrier concentrations in the junction region will rise to a very high value. As a result, die junction region contains a large concentration of electrons within the conduction band and *simultaneously* a large number of holes within the valence band. Holes represent absence of electrons. Thus, the upper energy levels in the narrow region are having a high electron population while the lower energy levels in the same region are vacant. Therefore, the condition of population inversion is attained in the narrow junction region. This narrow zone in which population inversion occurs is called an *inversion region* or *active region*. Chance recombination acts of electron and hole pairs lead to emission of spontaneous photons.

Q-SWITCHING

Since a laser is an oscillator, its resonator cavity is characterized by the quality factor, Q as is done in the case of an electronic oscillator. Q is defined as the ration of the energy W stored in the system to the energy losses AW per cycle. Thus,

$$Q = 2\pi \frac{\text{Energy stored in the resonator}}{\text{Energy lost in a cycle}}$$

In case of lasers Q is very high and is of the order of 108. Q-factor is a measure of the mirror losses.

The power of a laser beam can be drastically increased, provided a very large number of atoms in the active medium participates in stimulated emission. It can happen only when a very high population inversion density is established in the active medium. Normally when the laser starts oscillations, the high population inversion drops back to the threshold value of the steady state condition. It becomes therefore necessary that the onset of oscillations is delayed until largest possible number of atoms accumulates at the upper lasing level. The laser can be prevented from oscillating if, for example, the parallelism of the resonator mirrors is disturbed. If one of the end mirrors is misaligned, it cannot reflect incident photons into the active medium and therefore, stimulated emission cannot take place.

Consequently, the pumping process can build up the population inversion to a very high value in the medium. In effect, Q-factor of resonator is spoiled and optical losses are increased to a high value. If now the end mirror is aligned suddenly, it reflects photons into the medium. The feed back of photons triggers a chain of stimulated emissions and builds up rapidly a photon avalanche. Thus, laser oscillations set in suddenly and the Q of the cavity is increased abruptly. All the energy stored in the cavity is emitted in a single giant pulse with peak power much higher than the laser could produce otherwise. The pulse lasts for a short time and depopulates the upper energy level.quickly and the lasing action stops. This method of controlling the laser output power is called Q-switching method.

METHOD OF Q-SWITCHING

An electro-optic shutter can serve as a voltage controlled gate which rapidly switches the cavity from a passive to an active state. The shutter consists of a crystal that becomes double refracting when an electric field is applied across the crystal. The arrangement is shown in Fig. 3.20. A voltage is applied to the crystal during the pumping of the laser by the light from a flash lamp. The magnitude of the voltage is chosen such that it transforms the electro-optic crystal into a quarter- wave plate. Light emitted by the laser becomes linearly polarized on passing through the polarizer. The linearly polarized light is then incident on the electro-optic crystal and gets split into two mutually orthogonal components. As the two components travel through the electro-optic crystal, a relative phase retardation of 90° is produced between them. On emerging from the electro-optic crystal, the two components combine to produce circularly polarized light. The light beam reflects from the mirror and returns to the cavity travelling in the opposite direction. On reflection the sense of rotation of the circularly polarized light reverses and on repassing through the crystal, the two components of circularly polarized light experience a further retardation of 90° with respect to each other. Coming out of the crystal at the other end, the components recombine to produce linearly polarized light. The direction of polarization of the light is now al 90° with respect to its original direction of polarization and transmission axis of the polarizer. This light is not allowed to pass through the polarizer. Therefore, light does not come back to the laser rod and the cavity is switched off.

Thus, the cavity Q is reduced to a low value. When the voltage applied to the electro-optic crystal is turned off, double refraction is absent in the crystal and the state of polarization of the light passing through the crystal is unaffected. Light freely travels in both directions and returns to the laser rod to be further amplified. Now, the cavity is switched on and the Q regains its high value. The Q-switching is synchronized with pumping mechanism such at hat the voltage applied to the electro-optic crystal drops to zero value at the time when the population inversion in the laser medium attains its peak value.

LASER BEAM CHARACTERISTICS

The important characteristics of a laser beam are:

(i) directionality

(ii) negligible divergence

(iii) high intensity

(iv) high degree of coherence and

(v) high monochromaticity.

(i) *Directionality* : The conventional light sources emit light uniformly in all directions. When we need a narrow beam in a specific direction, we obtain it by placing a slit in front of the source of light.

In case of laser, the active material is in a cylindrical resonant cavity. Any light that is travelling in a direction other than parallel to the cavity axis is eliminated and only the light that is travelling parallel to the axis is selected and reinforced. Light propagating along the axial direction emerges from the cavity and becomes the laser beam. Thus, a laser emits light only in one direction.

(ii) *Divergence* : Light from conventional sources spreads out in the form of spherical wave fronts and hence it is highly divergent.

On the other hand, light from a laser propagates in the form of plane waves. The light beam remains essentially a bundle of parallel rays. The small divergence that exists is due to the diffraction of the beam at the exit mirror. A typical value of divergence of a He-Ne laser is 10^{-3} radians. It means that the diameter of the laser beam increases by about 1 mm for every meter it travels.

(iii) *Intensity* : The intensity of light from a conventional source decreases rapidly with distance as it spreads out in space. Laser emits light in the form of a narrow beam with its energy concentrated in a small region of space. Therefore, the beam intensity would be tremendously large and stays constant with distance. The intensity of a laser beam is approximately given by

$$I = \left[\frac{10}{\lambda}\right]^2 P \quad ...(1)$$

where P is the power radiated by the laser.

To obtain light of same intensity from a tungsten bulb, it would have to be raised to a temperature of 4.6×10^6K.

(iv) *Coherence* : The light that emerges from a conventional light source is a jumble of short wave trains which combine with each other in a random manner. The resultant light is incoherent. Coherence length is one of the parameters used as a measure of coherence. In case of a laser a large number of identical photons are emitted through stimulated emissions and therefore they will be in phase with each other. The resultant light exhibits a high degree of coherence. The coherence length of light from a sodium lamp, which is a traditional monochromatic source, is of the order of 0.3 mm. On the other hand the coherence length of light emitted by an ordinary helium-neon laser is about 100 m.

(v) *Monochromaticity* : If light coming from a source has only one frequency (single wavelength) of oscillation, the light is said to be *monochromatic* and the source a *monochromatic source.*

Light from traditional monochromatic sources spreads over a wavelength range of 100Å to 1000Å. On the other hand, the light from lasers is highly monochromatic and contains a very narrow range of a few angstroms (< 10Å).

LIGHT AMPLIFICATION

If we consider a medium in thermal equilibrium, there would be more atoms in the lower level than at higher level. That is $N_1 >> N_2$. As the probability for absorption transition is equal to the probability for stimulated transition, a photon travelling through the medium is more likely to get absorbed than to stimulate an excited atom to emit a photon. Therefore, usually the process of absorption dominates the process of stimulated emission. Similarly, an atom that is at the excited state is more likely to jump to the lower level on its own than being stimulated by a photon. Further, the photon density in the incident beam is not sufficient to interact with the excited atoms. Owing to this, the spontaneous emission dominates the stimulated emission.

Light amplification requires that stimulated emission occur almost exclusively. In practice, absorption and spontaneous emission always occur together with stimulated emission. The laser operation is achieved when stimulated emission exceeds in a large way the other two processes. Let us now look at the conditions under which the number of stimulated transitions can be made larger than the other two transitions.

Condition for Stimulated Emission to Dominate Spontaneous Emission

The ratio of eqn. (7) to eqn. (4) gives

$$R_1 = \frac{\text{Stimulated transitions}}{\text{Spontaneous transitions}} = \frac{B_{21}\rho(v)\, N_2}{A_{21}\, N_2} = \frac{B_{21}}{A_{21}}\rho(v)$$

The stimulated transitions will dominate the spontaneous transitions if the radiation density $\rho(v)$ is very large and the value of the ratio B_{21}/A_{21} is also large.

We have

$$R_1 = \left(\frac{B_{21}}{A_{21}}\right)\left[\frac{8\pi h v^3 \mu^2}{c^3} \cdot \frac{1}{e^{hv/kt} - 1}\right]$$

But

$$\frac{B_{21}}{A_{21}} = \frac{c^3}{8\pi h v^3 \mu^3}$$

$\therefore$

$$R_1 = \left(\frac{c^3}{8\pi h v^3 \mu^3}\right)\left[\frac{8\pi h v^3 \mu^3}{c^3} \cdot \frac{1}{e^{hv/kt} - 1}\right]$$

or

$$R_1 = \left[\frac{1}{e^{hv/kt} - 1}\right] \qquad ...(1)$$

If we assume $v = 5 \times 10^{14}$Hz

and $T = 300$ K, the value of R_1 comes to 10^{-58}.

The above result shows that in the optical region stimulated emission is negligible compared to spontaneous emission.

Condition for Stimulated Emission to Dominate Absorption Transitions

It may be noted that the presence of a large number of photons will lead to more absorption transitions rather than stimulated emissions. Hence large photon density alone will not guarantee more stimulated emissions.

$$R_2 = \frac{\text{Stimulated transition}}{\text{Absorption transition}} = \frac{B_{21}\,\rho(v)\,N_2}{B_{12}\,\rho(v)\,N_1} \qquad ...(2)$$

As $B_{21} = B_{12}$, $R_2 = \dfrac{N_2}{N_1}$ ...(3)

At thermodynamic equilibrium, $N_2 << N_1$ and the population at the ground level far exceeds, that of the excited level. When light propagates through the medium, a photon may hit an excited atom leading to stimulated emission, or be absorbed on hitting an atom in the ground state. As more number of atoms are there at the ground level, a photon has a much higher probability of being absorbed than of stimulating an atom at the excited level. Therefore, the absorption transitions will be larger than stimulated transitions and the medium will absorb the incident light. If, on the other hand, the number of atoms are more in the excited state, the opposite situation prevails and photons are more likely to cause stimulated transitions than absorption transitions.

Two sum up, three conditions are to be satisfied to make stimulated transitions overwhelm the other transitions:

(i) the population at excited level should be greater than that at the lower energy level,

(ii) the ratio $\dfrac{B_{21}}{A_{21}}$ should be large and

(iii) a very high radiation density should be present in the medium. A medium amplifies light only when these three conditions are fulfilled.

To achieve high percentage of stimulated emissions:

(i) an artificial situation known as *population inversion* is to be created in the medium,

(ii) A larger value of $\frac{B_{21}}{A_{21}}$ is achieved by choosing a metastable energy level as the higher level. As spontaneous transitions are forbidden from a *metastable state* the coefficient A_{21} will be smaller and as a result the ratio $\frac{B_{21}}{A_{21}}$ will be larger,

(iii) $\rho(v)$ is made larger by enclosing the emitted radiation in an *optical resonant cavity* formed by two parallel mirrors. The radiation is reflected many times till the photon density reaches a very high value and a favourable condition is created for large stimulated emissions.

AXIAL MODES

The wave properties of photons require that the waves be in phase and interfere constructively within the optical resonator such that progressive enhancement of light intensity takes place. The condition for constructive interference is that the path length travelled by a wave between two consecutive reflections at an end mirror should equal an integral multiple of the wavelength. It means that

$$2L = m\lambda \qquad (m = 1, 2, 3 \ldots)$$

or

$$L = m\,(\lambda/2) \qquad \ldots(1)$$

The above equations indicate that only those waves, which can form standing wave pattern can exist inside the cavity in a steady state. Waves of other wavelengths interfere destructively and are quickly attenuated.

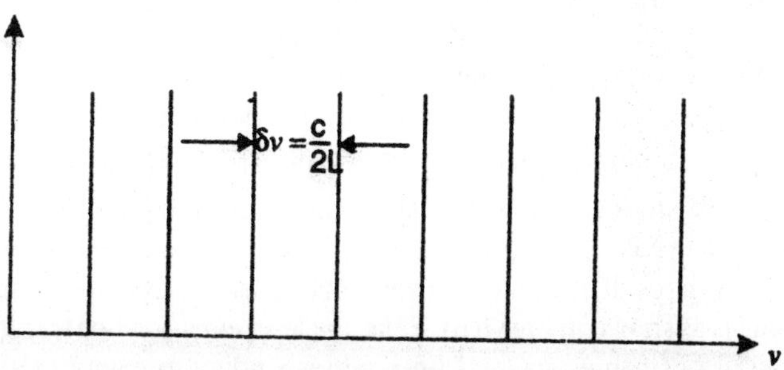

Fig. 6 : Cavity resonance frequencies representing the possible v longitudinal modes.

Because of its length which is very large compared to light wavelength, optical resonator supports simultaneously several standing waves of multiple wavelengths. These wavelengths are called *longitudinal* or *axial modes*. Therefore, m in eqn. (1) is called the *mode number*.

The frequencies of the modes are given by

$$\nu_m = \frac{mc}{2L} \qquad \text{...(2)}$$

Theoretically, the cavity can resonate at a very large number of frequencies (Fig. 6) that satisfy the equation (2). For example, if we take L = 0.5 m, and λ, = 5000Å, we obtain $m = 2 \times 10^6$. It means that the cavity supports 2×10^6 longitudinal modes.

GAIN CURVE AND LASER OPERATING FREQUENCIES

When the active medium is placed between the two mirrors, the cavity becomes an *active cavity*. Then, the standing waves supported by the cavity are the light waves emitted by the stimulated atoms of the medium. Ideally, the emission should occur at a single frequency, as a group of identical atoms radiate at the same frequency. However, in practice, because of the various line broadening mechanisms, there will be a spread of frequencies about the central frequency of the emission line. Fig. 7(a) shows the shape of a typical emission line. It is also called gain *curve* or *gain profile* because it indicates the range of frequencies over which stimulated emission can provide sufficient gain.

The laser operating frequencies are determined together by the resonant frequencies of the cavity and by the laser emission line width. If an output has to exist at a particular frequency, the cavity must be resonant at that specific frequency and there must be sufficient gain at that frequency. Laser oscillation can take place only when the gain is large enough to maintain resonance. For example, two dashed lines are shown in Fig. 7(a) which correspond to two threshold levels.

If the threshold is at level (1), then only the central frequency ν_0 is amplified. When the threshold is at (2), all frequencies between ν_1 and ν_2 can be amplified. However, there are only a few frequencies that resonate. As a result, the output of a laser consists of a few closely spaced frequencies as shown in Fig. 7(b). Thus, the laser emission line transforms into a series of names spectral lines corresponding to cavity modes. If δλ is the line width of the emission line, then me number of modes that would be ultimately present is given by

$$N = \frac{\delta\lambda}{\Delta\lambda} \quad ...(1)$$

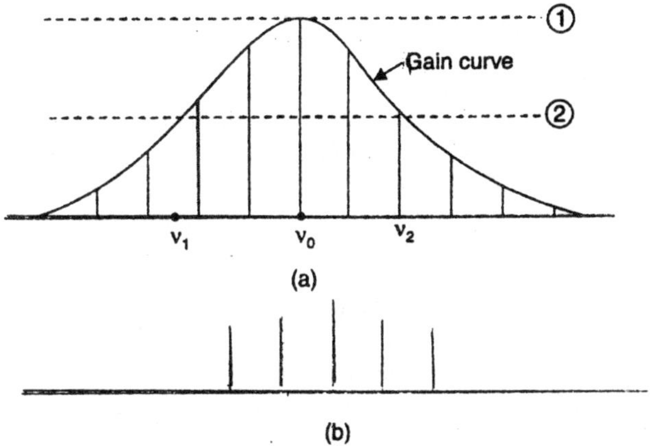

Fig. 7

TRANSVERSE MODES

The longitudinal dimension L of the resonant cavity governs the axial modes. The laser modes governed by the cross-sectional dimension of the optical cavity are called transverse *electromagnetic (TEM) modes*. The TEM modes are generally few in numbers and they are easy to see. If the laser beam is spread out by a negative lens and focussed on to a screen, several bright patches are seen on the screen. The patches are separated by intervals called nodal lines.

The transverse modes characterize the intensity distribution across the cross-section of the laser beam. In general, the allowed modes are designated as TEM_{mn}, where m and n are integers. The integers m and n represent the number of intensity minima in two orthogonal directions of the laser beam. The lowest order transverse mode is TEM_{00}. It is the simplest mode and has a smooth crosssection profile with a peak in the middle. TEM_{01}. beam has a single minimum dividing the beam into two bright spots. A TEM_{11} beam has two perpendicular minima dividing the beam into four quadrants, and so on. Operation of a laser in multimode form provides considerably more power than in single mode operation.

ATOM LASER

An atom laser is analogous to an optical laser. Optical laser generates a coherent beam of electromagnetic waves whereas an atom laser produces a coherent beam of matter waves. Each atom is associated with a de Broglie wave. When a dilute gas is cooled to temperatures near to zero, the de Broglie waves merge together to form a coherent matter wave. Therefore, an atom laser requires a thermal cloud of ultra-cold atoms which constitutes active medium of the laser. Hence, the main challenge in making an atom laser is the cooling of atoms to near zero temperature. In 1975 T.W. Hansch and A.L. Schawlow, USA, proposed first the technique of laser cooling. They suggested that neutral atoms might be cooled to millikelvin temperatures by the action of a near resonant laser beam that excites the atoms followed by spontaneous emission.

Steven Chu, Claude Cohen-Tannoudji and William D.Phillips succeeded in cooling and trapping atoms with laser light and were awarded in 1997 the Nobel prize in Physics for their work. Cooling to millikelvin temperatures is not enough to create coherent matter waves. Atoms are to be cooled to nanokelvin temperatures, when they form a Bose-Einstein condensate and create the state of coherent matter waves. In 1995 Comell, Wieman and their colleagues succeeded for the first time in cooling a dilute gas of rubidium atoms (Rb^{87}) to nanokelvin temperatures. They were awarded the Nobel prize in physics in the year 2001 for this work. The discovery of Bose-Einstein condensation has heralded the invention of atom laser. An atom laser will have a major impact on the fields of atom optics, atom lithography and precision measurements.

EVAPORATIVE COOLING

The atoms are further cooled to ultracold temperatures using the technique of evaporative cooling. The basic physics of evaporating cooling is very simple. It is known that a cup of hot milk or coffee cools down by evaporation. In the same way, high-energy atoms are allowed to escape from the sample so that the average energy of the remaining atoms is reduced.

It was shown by H.F. Hess that the idea of evaporative cooling could be applied to the atoms confined in a magnetic bowl. In this method the atoms are cooled first in the presence of the magneto-optic trap. Then the laser beams are turned off. Then a magnetic field, of the same type but much stronger than that used in MOT, is applied to perform the role of a bowl and confine the cold atoms. Now, a radio frequency oscillating field is applied to induce transitions in the atom between spin up state (attracted to the magnetic trap) and the spin down state (repelled by the magnetic trap), so that the atoms will undergo a spin-flip transition. If the rf field is tuned to the higher side of the magnetic bowl the atoms having higher energy will jump the well and fall. Lowering the rf frequency will induce some more atoms to leave the bowl. When the atoms with higher energy leave the bowl, the average energy of the remaining atoms becomes lower and atoms get colder in the process.

Continuing to cool involves continuously lowering the sides of the magnetic trap. The process is halted when the temperature goes down to nearly 100 nK. and only a few atoms are left to form a condensate, with the only drawback being that the number of trapped atoms is reduced. The process of Bose-Einstein condensate formation.

Bose-Einstein Condensate

Each atom has an associated de Broglie wavelength λ_B. In accordance with the Heisenberg uncertainty principle, the position of an atom is smeared out over a distance given by the thermal de Broglie wavelength,

$$\lambda_B = \left[\frac{h^2}{2kmT}\right]^{1/2}$$

where k is the Boltzmann constant, m is the atomic mass and T is the temperature of the gas. At room temperature the de Broglie wavelength is typically about ten thousand times smaller than the average distance between the atoms. This means that the matter waves of the individual atoms are uncorrelated or "disordered" and the gas can thus be described by *classical Boltzmann statistics*. As the gas is cooled, however, the wavelength increases, the smearing increases, and eventually there is more than one atom in each cube of dimension λ_B. The wave functions of adjacent atoms then "overlap", causing the atoms to lose their identity, and become one *'super-atom'*.

Bose-Einstein statistics dramatically increase the chances of finding more than one atom in the same state. Thus, at very low temperatures, all the atoms fall down to the lowest state, as they do not have enough energy to go to higher energy states. The result is *Bose-Einstein condensation*, a macroscopic occupation of the ground state of the gas. Although small, typically 0.1 mm across, a Bose-Einstein condensate can be seen with weakly magnifying lenses and a video camera. Thus, it is a microscopic object. The transition to BEC corresponds to a transition from a set of disordered atoms to coherent matter waves. Indeed, the transition from disordered to coherent matter waves can be compared to the change from incoherent to laser light. In a Bose condensate all the atoms occupy the same quantum state and can be described by the same wave function. The condensate therefore has many unusual properties not found in other states of matter.

BASIC ATOM LASER

An optical laser consists of an active medium held in a cavity resonator (Fig. 8). When the active medium is pumped, say with an optical source, into the state of population inversion, stimulated emission is triggered and an intense beam of light emerges out of the partially reflecting mirror of the cavity. An atom laser also should consist of similar principal parts.

1. *Active medium* : In the case of atom laser, the active medium is a thermal cloud of ultracold atoms.

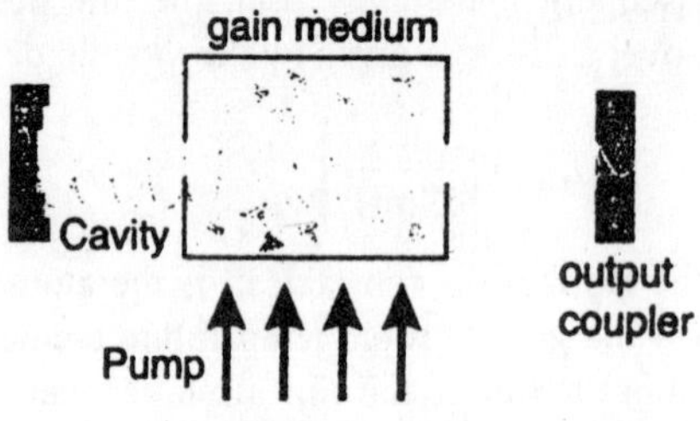

Fig. 8

2. *Cavity* : The cavity resonator is a magnetic trap in which the atoms are confined by magnetic mirrors. In a magnetic trap, for instance, once the atoms have been cooled and trapped by lasers, the light is switched off and an inhomogeneous magnetic field provides a confining potential around the atoms. The trap is analogous to the optical cavity formed by the mirrors in a conventional laser.
3. *Pumping* : Pumping is done by the evaporative cooling. The evaporation process creates a cloud which is not in thermal equilibrium and relaxes toward colder temperatures. This results in growth of the condensate.
4. *Stimulated emission* : In an atom laser, the presence of a Bose-Einstein condensate with N atoms enhances the probability that

an atom joins the condensate increasing the size of the condensate to N+1 atoms. The process of condensing atoms into the ground state of a magnetic trap is analogous to stimulated emission into a single mode of an optical laser.

5. *Laser threshold :* An optical laser starts working when the losses in the cavity are balanced by the gain in the medium. In atom laser the critical temperature for Bose-Einstein condensation resembles to the laser threshold. When the critical temperature for BEC is reached, atoms predominantly go into the lowest energy state of the system.

6. *Output-coupler :* An important feature of a laser is an output coupler to extract a fraction of the coherent field in a controlled way. In the case of a conventional laser the output coupler is a partially transmitting mirror.

We may think of the Bose condensate as being held in a container much like water in a bowl and to get a beam of coherent matter we just puncture the container and Bose condensate leaks out. The extraction process is called *out-coupling.*

Output coupling for atoms can be achieved by transferring them from states that are confined to states that are not, typically by changing an internal degree of freedom, such as the magnetic states of the atoms. Only the atoms that had their magnetic moments pointing in the opposite direction to the magnetic field were trapped. The "reflectivity" of the magnetic mirrors is altered by applying a short radio-frequency pulse to "flip" the spins of some of the atoms and therefore release them from the trap (Fig. 8). The extracted atoms then are accelerated away from the trap under the force of gravity. By changing the amplitude of the radio-frequency field, the extracted fraction could be varied between 0% and 100%. A pulsed output beam of atoms.

7. *Output of laser :* The output of an optical laser is a well collimated beam of light (electromagnetic waves). For an atom laser, the output is a beam of atoms (matter waves).

8. *Nature of output :* Optical lasers can work either in a pulsed mode or continuous wave (cw) mode. So far, the atom laser has been realized in the pulsed mode only. Most continuous-wave optical lasers are truly continuous in the-sense that they are continually fed energy or "pumped" so that they can supply photons indefinitely. A truly continuous source of coherent matter

waves could be similarly achieved only if the condensate could be replenished continually. Schemes for steady- state condensate formation are being explored currently.

9. *Modes :* Optical lasers radiate in several modes, *i.e.*, at several nearby frequencies. The cavity in an optical laser is typically many wavelengths long and as a result can support several different frequencies or modes. Atom laser operates in a single mode, since the Bose condensate actually involves mode competition and the atoms typically occupy the lowest energy state of the trap.

Important Features of a Laser Beam

1. *Monochromaticity :* An important property of laser light is that it is monochromatic. By analogy, in a Bose-Einstein condensate all the atoms have the same energy and hence the same de Broglie wavelength. If this property can be maintained when the atoms are released from the condensate, we will have a highly *monochromatic* source of matter waves.
2. *Coherence :* A crucial feature of a laser is the coherence of its output-in other words, the presence of a macroscopic wave.

   It is likely that the *coherence length* of the lasers is in fact much larger than the size of the condensate.
3. *Intensity :* Light produced by an optical laser is highly intense due to the constructive interference of coherent waves. Coherent matter waves are more intense than ordinary atoms.
4. *Directionality :* The spread of optical beam issuing out of a laser is governed by the diffraction effect at the output aperture. In case of atom laser, the beam is limited by Heisenberg uncertainty.

Differences Between an Atom Laser and an Optical Laser

1. In optical lasers population inversion is essential for realizing laser action. Population inversion does not occur in an atom laser.
2. In case of an optical laser, a large number of photons are generated through stimulated emission. In case of atom laser, atoms are not created. The atoms in the lower state are increased while the number of atoms in the upper states decreases.

3. Atoms are heavier than photns. They are therefore accelerated by gravity and a matter wave beam will fall like a beam of ordinary atoms.
4. A light beam from an optical laser can travel very far and does not require a vacuum for its operation. Unlike photons, atoms will not travel very far in air, so that the atom laser must be used in a vacuum.

ATOM LASER APPLICATIONS

The possibility of producing a coherent beam of atoms that could be collimated to travel long distances, or brought to a tiny focus like an optical laser, opens up a whole host of applications. Atom lasers may have a major impact on the fields of atom optics, atom lithography, precision atomic clocks and other measurements of fundamental standards. We study here three typical applications of atom laser.

Holography

One application for which the coherence of an atom laser is critical is atom holography. Just as conventional holography uses the diffraction of a photon beam to reconstruct a 3-D image, atom holography uses the diffraction of atoms. As the de Broglie wavelength of the atoms is much smaller than the wavelength of light, an atom laser could create much higher resolution holographic images. Atom holography might be used to project complex integrated-circuit patterns, just a few nanometres in scale, onto semiconductors.

The first atom holograms were demonstrated in 1996 by Fujio Shimuzu and colleagues at the University of Tokyo, using laser-cooled atoms. In the case of laser-cooled gases, the level of coherence needed to create a hologram is achieved by selecting a small portion of the atoms. The problem would be simplified if a source, in which most of the atoms are in the same quantum state, is used. An atom laser is such a source and could provide a much more intense and fully coherent beam of atoms.

Holography is a two-step process. First, a hologram—a sort of diffraction grating containing information about the object—is produced. Then a beam of light (or atoms) is diffracted by the hologram to form the image. In optical holography, the hologram is often made by interfering a laser beam with light that has been reflected from an object. The resulting diffraction pattern is recorded on photographic film.

In the atom-holography experiments, an image has been created by diffracting a coherent beam of atoms through a grating that was manufactured using electron-beam lithography. The image may be recorded on a "microchannel plate"—a detector that is sensitive to atoms. So far, atom holography has been able to produce 2-D images.

Atom Interferometry

Another important application is atom interferometry. In an atom interferometer an atomic wave packet is coherently split into two wave packets that follow different paths before recombining. The interference pattern created when the two wave packets recombine tells us something about the phase difference between the two paths. Atom interferometers that are more sensitive than optical interferometers could be used to test quantum theory, and may even be able to detect changes in space-time. This is because the de Broglie wavelength of the atoms is smaller than the wavelength of light, and the atoms have mass. Atom lasers would allow the use of devices with unequal path lengths, such as Michelson interferometers. Such devices may provide a way to measure lengths over large distances with unprecedented precision.

Nonlinear Atom Optics

Just as the invention of the laser enabled the field of nonlinear optics to flourish, intense sources of coherent matter waves have opened up a similar field in atom optics. Until recently, most atom-optics experiments could be thought of as single-particle phenomena where the interactions between particles could be neglected. In conventional nonlinear optics, photons interact with each other through some mediating material, such as a transparent crystal. A common nonlinear optical phenomenon is "four-wave mixing". Typically three waves, of frequency ω_1, ω_2 and ω_3, are sent into a nonlinear crystal. The exchange of energy and momentum between the waves, mediated by the nonlinear crystal, results in the production of a fourth wave with frequency $\omega_4 = \omega_1 + \omega_2 - \omega_3$ (Fig. 9.13a). A quantum mechanical description of this process shows that two photons from separate beams annihilate in the crystal and produce two new photons. The energy and momentum of one of these photons adds to the third beam, while the other photon corresponds to a new, fourth beam.

In 1998 an analogous process with matter waves was predicted. It is predicted that if three condensates of appropriate momenta collided, the term in the nonlinear Schrodinger equation that describes the interactions between the atoms would give rise to a fourth. At the atomic level, this process can be described as a collision between two atoms from separate matter-wave beams. One of the atoms is stimulated so that it scatters in the direction of the third incident matter-wave beam. By the conservation of momentum, the other atom goes off to make a fourth, separate beam.

The actual experiment did not use three separate condensates. Instead, lasers were used to divide H single condensiale into three different momentum states via a process called Bragg diffraction. Starting with a condensiale at rest, two separate pulses of interfering laser beams were applied to create the Bragg "diffraction grating" that divided the atoms roughly equally into three different momentum states, including the state of the initial condensate. When these pulses were applied fast enough-that is Before the different momentum states had a chance to separate-atoms in a fourth momentum state were produced.

The four-wave mixing process arises from collisions between pairs of atoms from two matter- wave beams (1 and 2 in). One pair of atoms scatters in the direction of the third, incident Biatter-waye beam and-amplifies (3). By the conservation of energy and momentum, the other pair of atoms produces a fourth, separate beam (4). Fig. 3.28(c) shows an image of the experimental atomic distribution showing the fourth (small) wave packet generated by the matter-wave mixing process.

BOSE-EINSTEIN CONDENSATION

It is necessary to understand what a Bose-Einstein condensate is and how it is produced, before we go into the details of atom laser.

In 1924 the Indian physicist Satyendra Nath Bose derived the Planck law for black-body radiation by treating the photons as a gas of identical particles. Photons have an intrinsic angular momentum, or "spin", of the Planck constant h divided by 2π. Einstein generalized Bose's theory to other particles of integral spin. The theory is now known as *Bose-Einstein statistics*. Particles that have a spin that is an integer multiple of h/2π obey Bose-Einstein statistics and are called bosons.

It was shown that more than one boson can occupy the same quantum state. Einstein predicted that at sufficiently low temperatures all the atoms in an ideal gas of identical atoms might be condensed to a single lowest quantum state of the system. This large number of atoms locked together in the same state is called *Bose-Einstein condensate* (BEC). Bose-Einstein condensation is an exotic quantum phenomenon that was observed in dilute atomic gases. Bose-Einstein condensation happens only for "bosons". The low temperature necessary for obtaining BEC is of the order of nanokelvin.

METHODS OF COOLING ATOMS

Let us consider a gas inside a container at room temperature. The atoms of gas move at random in all directions and attain thermal equilibrium with the container. They move with velocities, υ, which are of the order of a few hundred meters/second. According to the kinetic theory the absolute temperature is proportional to $\left\langle \upsilon^2 \right\rangle$. Temperature is an expression of kinetic motion of the particles. Hence the kinetic energy of a particle may define the kinetic temperature.

The successful route to form Bose-Einstein condensate of the gas involves two steps. The first step consists in cooling down the atoms at low density to millikelvin temperatures by the action of a laser beam of an appropriate frequency. In the second step, the cold atoms are further cooled by using the technique of evaporative cooling. These techniques provided a new route to ultracold temperatures that does not involve cryogenics.

LASER DOPPLER COOLING

The basic principle of this method is as follows. If a photon falls on an atom, which has a resonance frequency equal to that of the photon, the atom absorbs the photon. If an atom absorbs a photon coming from the opposite direction, the momentum of the photon is transferred to the atom and the atom is pushed back and loses its velocity. In order to ensure deceleration of the atom, the atoms should absorb only the oppositely moving radiation. This is achieved by using Doppler effect.

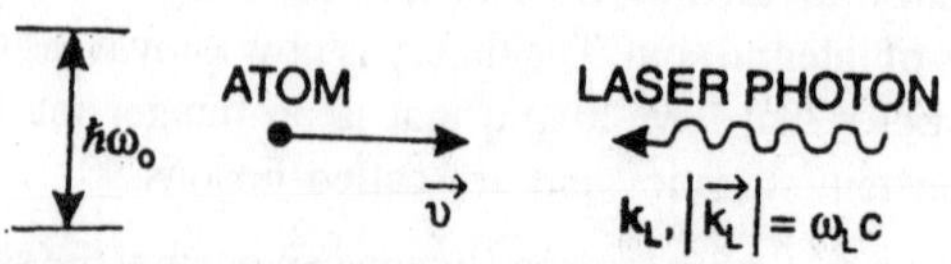

Fig. 9 : Doppler-resonant absorption of a red-detuned laser beam by a moving two-leve) atom

An atom with a lower energy level E_1 and an upper energy level E_2 loses or gains energy by emitting or absorbing light of frequency v_L such that $E_2 - E_1 = h\nu_L$. The photons associated with a plane light wave of frequency ν_L, propagating along a direction with unit vector k_L have an energy $E = h\nu_L$ and a momentum $P_L = \left(\frac{h\nu_L}{c}\right)$ each. Absorption of a single photon by an atom results in a momentum transfer of P_L to the atom. In one second, the atom can absorb and emit a large number of times and hence-the net momentum transfer rate to the tiny atom become substantial. After about ten thousand absorption and subsequent emission cycles, an atom which was originally moving at about 700 m/s can be slowed down to near zero speed. The laser frequency ν_L is selected such that it is just beneath the resonant frequency ν_0 of the atoms. Because of its motion, any particular atom sees an oncoming photon with a frequency that is Doppler-shifted upward by an amount $\nu_L \upsilon/c$. When the laser frequency is tuned so that $\nu_0 = \nu_L (1 + \upsilon/c)$, collisions with photons will resonate the atoms. In the process, each photon transfers its momentum of $\hbar k_L$ to the absorbing atom, whose speed there upon reduced by an amount $\Delta\upsilon$ where $m\,\Delta\upsilon = \hbar k_L$. The cloud of atoms is not very dense, and each excited atom can drop back to its ground state with spontaneous emission of a photon of energy $h\nu_0$. The emission is randomly directed and so although the atom recoils, the average amount of momentum regained by it over thousands of cycles tends to zero. The change in momentum of the atom per photon absorption-emission cycle is therefore effectively $\hbar k_L$ and it slows down. By contrast, an atom moving in the opposite direction, away from the light source sees photons to have a frequency $\nu_L (1 - \upsilon/c)$, far enough away from ν_0 that there can be little or no absorption and therefore no momentum gain. If there are two counter-propagating waves, they can be used to decelerate the atoms coming from both directions.

The process is repeated several times and slows down the atoms. Hence, the gas is cooled. Thus, the absorption slows down the atom where as the emission does not have an effect and on the average there is a reduction in momentum. However, the lowest temperature that the laser cooling technique can reach is limited. The process of Doppler cooling can lower the temperature to the order of 100 μK for alkali atoms. Alkali atoms are well suited to laser- based methods because their optical transitions can be excited by available lasers and because they have a favourable internal energy-level structure for cooling to low temperatures. Once the atoms are cooled to uX temperature, we have to confine them. If the cooling laser is turned off, the atoms will start moving apart with their residual velocities, and will hit the walls and eventually fall down. The cold atoms usually have a velocity of a few centimeters per second.

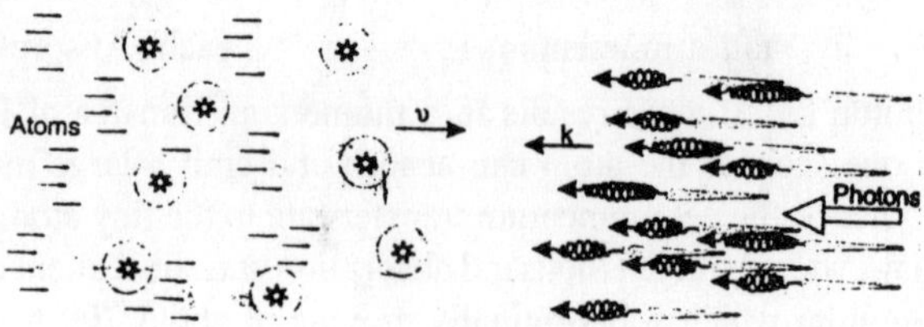

Fig. 9 : A stream of atoms colliding with a laserbeam in the process of laser cooling.

In practice six laser beams are arranged-two counter-propagating beams each, (one along the positive direction and the other along the negative direction) along each of the three Cartesian axes, as shown in Fig. 10. An atom, in whichever direction it moves, always encounters abeam in opposite direction and the net effect is that the atom experiences a strong viscous force in the three dimensional space in the field of photons.

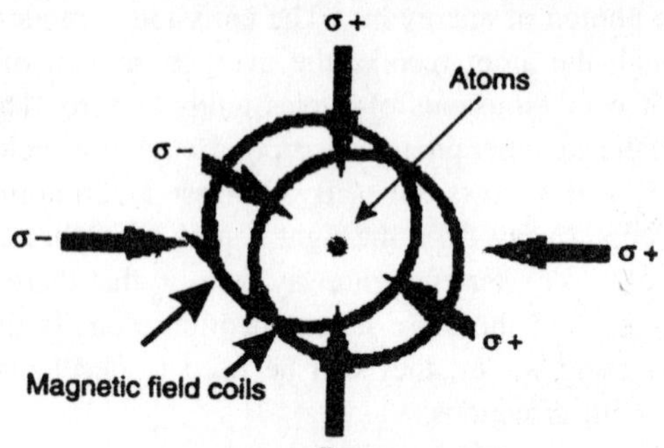

Fig. 10 : Six laser beams and a pair of magnetic field coils forming a magneto-optical trap can cool more than a billion atoms to micro-kelvin temperatures.

This six laser beam configuration used to cool a sample of atoms from a vapour is given the name *optical molasses*, since the light bath always opposes the motion of atoms as if they were submerged in molasses and causes viscous effect on the atomic motion. The optical molasses could produce a temperature much lower than Doppler cooling limit. However there is no position dependent force and since the atoms are not slowed to complete standstill, they diffuse out of the laser beams and fall away under gravity.

Trapping of Atoms

The frictional force arising from the Doppler effect cools the atoms, but does not capture them; the atoms fall out of the molasses in a few seconds. It becomes therefore necessary to confine the atoms in a small region of space. Position dependence is introduced by using ($\sigma^+/\sigma-$) polarized laser beams and applying a weak magnetic field. This arrangement is known as a *magneto-optical trap* (MOT). The trap may be a spherical quadrupole magnetic field, produced by two anti-Helmholtz coils carrying currents in opposite directions. Such an arrangement gives rise to a zero-field region between the two coils, at the intersection of the six laser beams and the field increases in all directions away from this point. This field minimum attract molecules in the low-field seeking states, that is molecules with their magnetic moments oriented anti-parallel to the magnetic fields, and repel strong field seeking molecules (Fig. 10).

Alkali atoms have a magnetic moment because they have an unpaired electron. The magnetic moment is in a direction opposite to that of the electron spin. The magnetic moment interacts with the applied weak magnetic field. If the magnetic moment is parallel to the external magnetic field, the atom is attracted to the local minimum of the field (Fig. 11) and can be trapped. It is also necessary that the atoms must be thermally isolated from their surroundings, since at ultracold temperatures atoms stick to all surfaces.

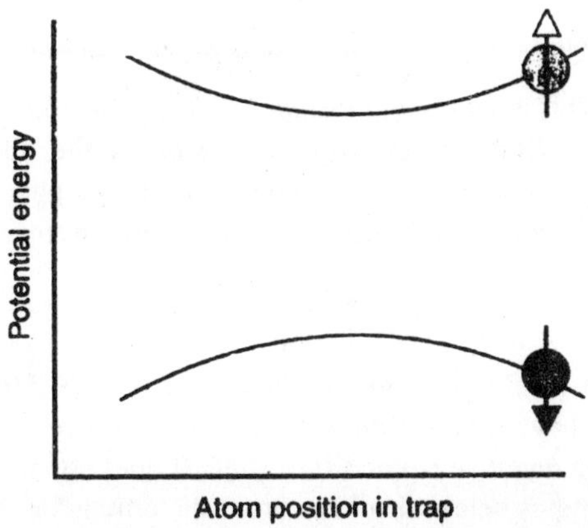

Fig. 11 : An atom with spin parallel to the magnetic field is attracted to the energy minimum and an atom with anti-parallel spin is repelled.

After the atoms are trapped and cooled with lasers, all light is extinguished and a potential is built up around the atoms with an inhomogeneous magnetic field. Thus, the magnetic field can act as a little bowl and which confines the atoms to a small region of space. In addition to holding the atoms at a point in space, the trapping force compresses them into a dense cold cloud.

POPULATION INVERSION

When the material is in thermal equilibrium condition, the population ratio is governed by the Boltzmann distribution law according to the following equation:

$$\frac{N_2}{N_1} = e^{-[E_2 - E_1]/kT} \quad ...(1)$$

It means that the population N_2 at the excited level E_2 will be far smaller than the population N_1 at the level E_1. The condition in which there are more atoms in the lower energy level and relatively lesser number of atoms in the higher energy level is called *normal condition* or *thermal equilibrium*. Thus, under thermal equilibrium,

$$N_1 >> N_2.$$

To achieve a high percentage of stimulated emission, a majority of atoms should be at the higher energy level than at the lower level. We may somehow enhance the number of atoms in the excited level, such that the population ratio $\frac{N_2}{N_1}$ *momentarily* increases without change in temperature. This is a *non-equilibrium condition* and is known as *inverted population* condition. *Population inversion* is the non-equilibrium condition of the material in which population of the upper energy level N_2 momentarily exceeds the population of the lower energy level N_1. That is,

$$N_2 >> N_1 \quad ...(2)$$

From eqn. (2) it is seen that N_2 can exceed N_1 only if the temperature were negative. In view of this, the state of population inversion is sometimes referred to as a *negative temperature state*. It does not mean that we can attain temperatures below absolute zero. The terminology underlines the fact that the state of population inversion is a non-equilibrium state. It should be borne in mind that the population inversion is attained at normal temperatures.

The system has two energy levels. At thermal equilibrium, photon absorption and emission processes take place side by side, but because $N_1 > N_2$ the system absorbs photons rather than emits photons. Now suppose that the system is supplied with energy from an external source till N_2 exceeds N_1. Then, the system is said to have attained the state of population inversion. The population inversion has taken place between the levels E_2 and E_1.

When the system is in the population inversion condition, a few randomly emitted photons trigger stimulated emission of photons and those stimulated photons induce more stimulated emissions and so on. Consequently light gets amplified (Fig. 12) and a cascade of light is produced. However, in this process atoms from E_2 level make downward transitions and as soon as the population at lower level becomes equal or larger than that at the excited level, population inversion comes to an end. Energy is again to be supplied to the system to take it into the state of population inversion.

The non-equilibrium condition (population inversion) is attained by employing pumping techniques to transfer large number of atoms from lower energy level to higher energy level.

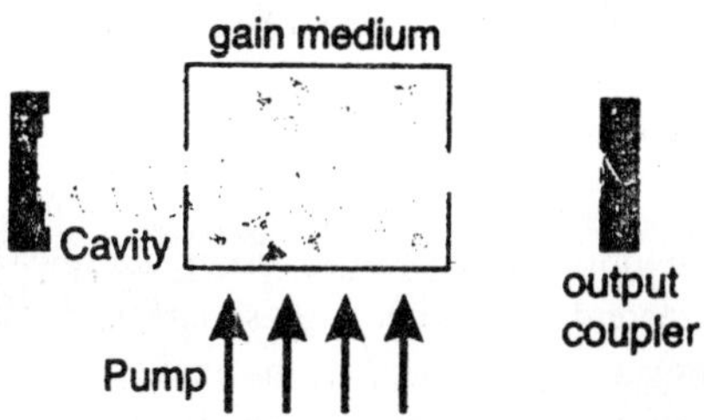

Fig. 12 : Amplification of a light wave in a medium with population inversion.

ACTIVE MEDIUM

Atoms are in general characterized by a large number of energy levels. However, all types of atoms are not suitable for laser operation. Even in a medium consisting of different species of atoms, only a small fraction of atoms of a particular type have energy level system suitable for achieving population inversion. Such atoms can produce more stimulated emission than spontaneous emission and cause amplification of light. Those atoms, which cause laser action, are called *active centers*.

The rest of the medium acts as host and supports active centers. The medium hosting the active centers is called the *active medium*. An *active medium is a medium which when excited reaches the state of population inversion and promotes stimulated emissions leading to light amplification.*

PUMPING

For achieving and maintaining the condition of population inversion, we have to raise continuously the atoms in the lower energy level to the upper energy level. It requires energy to be supplied to the system. *Pumping* is the process of supplying energy to the laser medium with a view to transfer it into the state of population inversion. Because N_1 is originally very much larger than N_2, a large amount of input energy is required to momentarily increase N_2 to a value comparable to N_1. There are a number of techniques for pumping a collection of atoms to an inverted state. Optical pumping, electrical discharge and direct conversion are some of the methods of pumping. In *optical pumping*, a light source such as a flash discharge tube is used to illuminate the active medium. This method is adopted in solid state lasers. In *electrical discharge* method, the electric field causes ionization of the medium and raises it to the excited state. In semiconductor diode lasers, a *direct conversion* of electrical energy into light energy takes place.

METASTABLE STATES

An atom can be excited to a higher level by supplying energy to it. Normally, excited atoms have short lifetimes and release their energy in a matter of nanoseconds (10^{-9}s) through spontaneous emission. It means that atoms do not stay long enough at the excited state to be stimulated. As a result, even though the pumping agent continuously raises the atoms to the excited level, they undergo spontaneous transitions and rapidly return to the lower energy level. Population inversion cannot be established under such circumstances. In order to establish the condition of population inversion, the excited atoms are required to 'wait' at the upper energy level till a large number of atoms accumulate at that level. In other words, it is necessary that the excited state has a longer lifetime. A metastable state is such a state. Because of restrictions imposed by conservation of angular momentum, an electron excited to a metastable state cannot return to the ground state by emitting a photon, as it is generally expected to do. Such a state in which single-photon emission is impossible, has an unusually long time and is called a *metastable state.* Atoms excited to the metastable states remain excited for an appreciable time, which is of the order of 10^{-6} to 10^{-3}s. This is 10^3 to 10^6 times the lifetimes of the ordinary energy levels.

Therefore, the metastable state allows accumulation of a large number of excited atoms at that level. The metastable state population can exceed the population at a lower level and establish the condition of population inversion in the lasing medium. It would be impossible to create the state of population inversion without a metastable state. Metastable state can be readily obtained in a crystal system containing impurity atoms. These levels lie in the forbidden band gap of the host crystal. Population inversion readily takes place as the lifetimes of these levels are large, and secondly, there is no competition in filling these levels, as they are localized levels.

There could be no population inversion and hence no laser action, if metastable states do not exist.

PRINCIPAL PUMPING SCHEMES

Atoms in general are characterized by a large number of energy levels. Among them only three or four levels will be pertinent to the pumping process. Therefore, only those levels are depicted in the pumping scheme diagrams. Two important pumping schemes are widely employed. They are known as three-level and four-level pumping schemes.

THREE-LEVEL PUMPING SCHEME

A typical three-level pumping scheme is shown in Fig. 13. The state E_1 is the ground level; E_3 is the pump level and E_2 is the metastable upper lasing level. When the medium is exposed to pump frequency radiation, a large number of atoms will be excited to E_3 level. However, they do not stay at that level but rapidly undergo downward transitions to the metastable level E_2 through non-radiative transitions. The atoms are trapped at this level as spontaneous transition from the level E_2 to the level E_1 is forbidden. The pumping continues and after a short time there will be a large accumulation of atoms at the level E_2. When more than half of the ground level atoms accumulate at E_2, the population inversion condition is achieved between the two levels E_1 and E_2. Now a chance photon can trigger stimulated emission.

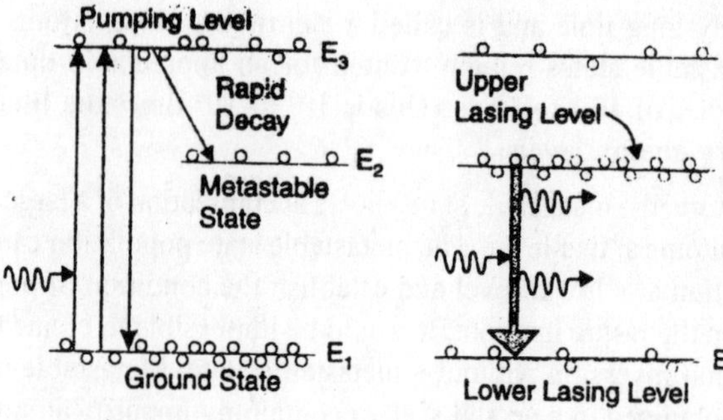

Fig. 13 : A typical three level pumping scheme—(a) optical pumping (fc) lasing action

Four-Level Pumping Scheme

A typical four-level pumping scheme is shown in Fig. 3.35. The level E_1 is the ground level, E_4 the pumping level, E_3 the metastable upper lasing level and E_2 the lower lasing level. E_2, E_3 and E_4 are the excited levels. When light of pump frequency v_ρ is incident on the lasing medium, the active centers are readily excited from the ground level to the pumping level E_4. The atoms stay at the E_4 level for only about 10^{-8}, and quickly drop down to the metastable level E_3. As spontaneous transitions from the level E_3 to level E_2 cannot take place, the atoms get trapped at the level E_3. The population at the level E_3 grows rapidly. The level E_2 is well above the ground level such that $(E_2 - E_1) > kT$. Therefore, at normal temperature atoms cannot jump to level E_2 from E_1 on the strength of thermal energy. As a result, the level E_2 is virtually empty. Therefore, population inversion is attained between the levels E_3 and E_2.

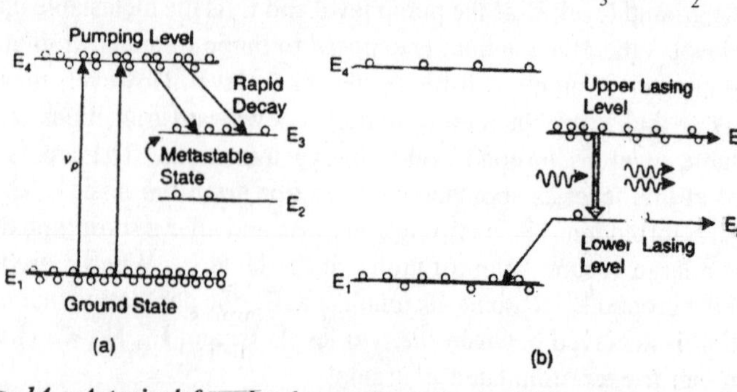

Fig. 14 : A typical four level pumping scheme (a) Pumping (b) Lasing action.

A chance photon of energy hv = $(E_3 - E_2)$ emitted spontaneously can start a chain of stimulated emissions, bringing the atoms to the lower laser level E_2. From the level E_2 the atoms subsequently under go non-radiative transitions to the ground level E_1 and will be once again available for excitation.

Comparison of Four-Level Laser with the Three-Level Laser

1. In the three-level pumping scheme, the terminal level of laser transition is simultaneously the ground level. Therefore, in order to achieve population inversion more than half of the ground level atoms have to be pumped up to the upper lasing level, such that $N_2 > N_1/2$. As the number of atoms in the ground level is very large, high pump power is required in order to promote $N_1/2$ atoms and establish the required population inversion. On the other hand, in the four-level pumping Scheme, the terminal level of laser transition is virtually empty and population inversion condition is readily established even if a smaller number of atoms arrive at the upper lasing level. Therefore, relatively small pumping power is required to establish population inversion in four level pumping schemes.
2. In case of three level pumping scheme, once stimulated emission commences, the population inversion condition reverts to normal population condition.. Lasing ceases as soon as the excited atoms drop to the ground level. Lasing occurs again only when the population inversion is re-established. The light output therefore is a *pulsed output*.

In case of four level scheme, the condition of population inversion can be held without interruption and light output is obtained continuously. Thus, the laser operates in *continuous wave* (cw) mode.

Necessity of Broad absorption Band at Pumping Level

In a laser medium, active centers are excited through absorption of energy from a pumping source. We desire that the energy given through the pumping agent is utilized to the largest possible extent in exciting the ground level atoms. This is denoted by *pumping efficiency*. In case of optical pumping, a flash discharge from a lamp serves as the pumping agent. Normally, light sources such as the flash discharge emit light over a wide frequency range. More number of atoms can be excited to the higher level if large number of the frequency components is utilized instead of a single frequency. That can happen *only* when there is a band of close spaced energy levels at the pumping level.

Thus, for a larger pumping efficiency, the pump level should be a broad band rather than a narrow discrete level.

OPTICAL RESONANT CAVITY

Laser is a light source and it is analogous to an electronic oscillator. An electronic oscillator (Fig. 15) is essentially an amplifier supplied' with a positive feed back. A part of the output of the amplifier is taken and fed back at its input. When the amplifier is switched on, electrical noise signal of appropriate frequency present at the input will be amplified; the output is fed back to the input and amplified again and so on. A stable output is quickly reached when the oscillator acts as a source of a particular frequency.

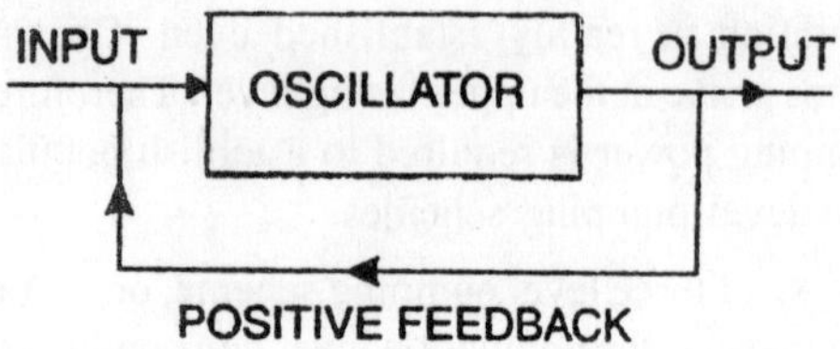

Fig. 15

In laser the active medium is the amplifying medium. It is converted into an oscillator through the feed back mechanism established by an optical resonator. A pair of optically plane parallel mirrors (Fig. 16) constitutes an *optical resonant cavity*. It is known as a *Fabry-Perot resonator*. One of these mirrors is fully reflecting and reflects all the light that is incident on it. The other mirror is made partially reflecting such that more than 90% of incident light is reflected from it and a small fraction is transmitted through it as the laser beam.

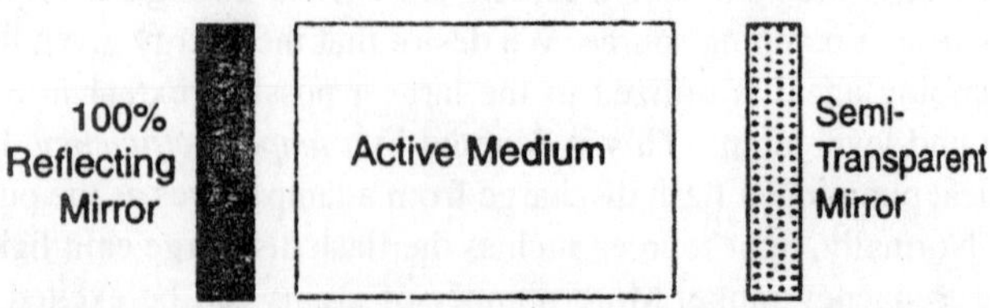

Fig. 16 : Fabry-Perot optical resonator

In laser, the role of noise is played by chance photons emitted spontaneously. The photons emitted along the optic axis of the resonant cavity travel through the medium and trigger stimulated emissions. They are reflected by the end mirror and reverse their path. The photons are thus fed back into the medium and travel toward the opposite end mirror causing more stimulated emissions. The photons are once more reflected at the mirror and travel toward the opposite mirror. Substantial light amplification takes place because the light beam is reflected several times at the mirrors and gains strength in each passage. Ultimately, when the amplification balances the losses in the cavity, the laser beam emerges out from the front—end mirror. *In the absence of resonator cavity, there would be no generation of light.*

Lasing Action

The active centers in the medium are in the ground state initially. Through suitable pumping mechanism, the medium is taken into the state of population inversion. A majority of photons travelling along the axis are reflected back on reaching the end mirror. They travel towards the opposite mirror and on their way stimulate more and more atoms and build up the photon strength. The photons that strike the opposite mirror are reflected once more into the medium. The photons travel once more through the medium generating more photons and more amplification. The photons are then reflected again at the mirror and travel through the medium. As the photons are reflected back and forth between the mirrors, stimulated emission sharply increases and the amplification of light is augmented. The mirrors thus provide positive feed back of light into the medium so that stimulated emission acts are sustained and the medium operates as an oscillator. At each reflection at the front-end mirror, light is partially transmitted through it. The transmitted component constitutes a *loss of energy* from the resonator. When the losses at the mirrors and within the medium balance the gain, the laser oscillations build up. A steady and strong laser beam will emerge from the front-end mirror. As the end mirrors reflect light strongly into the laser medium, the light levels are very high with in the optical cavity. Consequently, the condition that $\rho(\nu)$ should be large within the medium is satisfied.

THRESHOLD FOR OSCILLATION

As the light bounces back and forth in the optical resonator, it undergoes amplification as well as it suffers various losses. The losses occur mainly due to transmission at the output mirror and due to the scattering and diffraction of light within the active medium. For the proper build up of oscillations, it is essential that the amplification between two consecutive reflections of light from rear end mirror balance the losses. We can determine the threshold gain by considering the change in intensity of a beam of light undergoing a round trip within the resonator. Let us assume that the laser medium fills the space between the mirrors M_1 and M_2 (Fig. 17), which have reflectivity r_1 and r_2 respectively. Let the mirrors be separated by a distance L. Further, let the intensity of the light beam be I_o at M_1. Then, in travelling from mirror M_1 to mirror My the beam intensity increases from I_o to I(L), which is given by

$$I(L) = I_o e^{(\gamma - \alpha_s)L} \qquad ...(1)$$

After reflection at M_2 the beam intensity will be $r_2 I_o e^{(\gamma-\alpha_s)L}$ and after a complete round trip the final intensity will be

$$I\,(2L)\; r_1 r_2 I_o e^{(\gamma-\alpha_s)2L} \qquad ...(2)$$

The amplification obtained during the round trip is

$$G = \frac{I(2L)}{I_o} = r_1 r_2 e^{(\gamma-\alpha_s)2L} \qquad ...(3)$$

The product $r_1 r_2$ represents the losses at the mirrors whereas α_s includes all the distributed losses such as scattering, diffraction and absorption occurring in the medium. The losses are balanced by gain, when $G \geq 1$ or $I(2L) = I_o$. It requires that

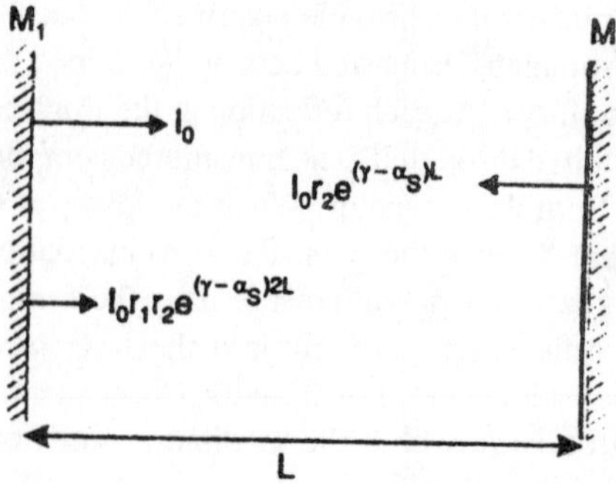

Fig. 17

$$r_1 r_2\, e^{2(\gamma-\alpha_s)L} \geq 1 \qquad ...(4)$$

or $$e^{2(\gamma-\alpha_s)L} \geq \frac{1}{r_1 r_2}$$

Taking logarithms on both sides, we get

$$2L(\gamma - \alpha_s) \geq \text{In } r_1 r_2$$

$$\gamma - \alpha_s \geq -\frac{1}{2L} \text{ In } r_1 r_2$$

$$\gamma \geq \alpha_s - \frac{1}{2L} \text{ In } r_1 r_2 \qquad ...(5)$$

or $$\gamma \geq \alpha_s + \frac{1}{2L} \text{ In } \frac{1}{r_1 r_2} \qquad ...(6)$$

Eqn. (6) is known as the condition for lasing. It shows that the initial gain must exceed the sum of the losses in the cavity. This condition is used to determine the threshold value of pumping energy for lasing action.

γ, the amplification of the laser will be dependent on how hard the laser medium is pumped. As the pump power is slowly increased, a value of γ_{th} called threshold value is reached and the laser starts oscillating. The threshold value γ_{th} is given by

$$\gamma_{th} = \alpha_s + \frac{1}{2L} \text{ In } \frac{1}{r_1 r_2} \qquad ...(7)$$

Eqn. (20) states the condition when the net gain would be able to counteract the effect of losses in the cavity and is known as the *threshold condition for lasing*. The value of γ must be atleast γ_{th} for laser oscillations to commence.

Function of Resonant Cavity

(i) The primary function of optical resonator is to provide a positive feedback of light into the lasing medium so that the stimulated emission acts are sustained and the laser acts as a generator of light.

(ii) A chance photon spontaneously emitted by an excited atom acts as the input and induces stimulated emission. To sustain stimulated emission acts and to increase the light intensity in a cumulative way, a positive feedback of light must be provided. The mirrors by way of reflecting the incident photons provide the feedback.

(iii) Laser oscillation is initiated by the photons spontaneously radiated by some of the excited atoms.

Each spontaneous photon can trigger many stimulated transitions along the path of its travel. As the initial spontaneous photons are emitted in various directions, the secondary stimulated photons will also travel in various directions. The result would be production of incoherent light. In order to build up oscillations, a specific direction has to be defined for photon propagation in the lasing medium. The optical resonator sets its optic axis as the most favourable direction for build-up of light beam.

(iv) In order to make the stimulated emissions dominate spontaneous emissions, a high optical energy density $\rho(v)$ is necessary to be present in the active medium. The mirrors constituting the cavity confine more than 90% of the emitted photons to be within the laser medium such that a very high optical energy density is always present in the lasing medium.

(v) The optical cavity is very much similar to a resonating column. Just as standing waves form in a resonating column, standing waves of optical frequencies are formed in the optical cavity. If L is the length of the cavity, the longest wavelength that will produce a standing wave pattern is $\lambda. = 2L$. In general the cavity supports the wavelengths

$$\lambda_m = \frac{2L}{m} \ (m = 1, 2, 3, 4 \ldots) \tag{8}$$

Waves of other wavelengths attenuate quickly. Thus, the optical cavity selects and amplifies only certain frequencies. Hence, the value of L should be properly selected.

(vi) Active centers may have a number of lasing transitions instead of only one transition. The mirrors of optical cavity suppress the undesired transitions. The reflectivity of the mirrors is further made less for the undesired photons, which therefore get absorbed at the mirrors.

HELIUM-NEON LASER

Gas lasers are the most widely used lasers. They range from the low power helium-neon laser used in college laboratories to very high power carbon dioxide laser used in industrial applications. These lasers operate with rarefied gases as the active media and are excited by an electric discharge. In gases, the energy levels of atoms involved in the lasing process are narrow and as such require sources with sharp wavelength to excite atoms. Finding an appropriate optical source for pumping poses a problem. Therefore optical pumping is not used in gas lasers. The most common method of exciting gas laser medium is by passing an electric discharge through the gas. Electrons present in the discharge transfer energy to atoms in the laser gas by collisions.

The first gas laser was He-Ne laser which was invented in 1961 by Ali Javan, William R. Bennett, Jr. and Donald R.Hemott.

Construction : The schematic of a He-Ne laser is shown in Fig. 3.5. Helium-Neon laser consists of a long discharge tube filled with a mixture of helium and neon gases in the ratio 10:1. Neon atoms are the active centers and have energy levels suitable for laser transitions while helium atoms help in exciting neon atoms. Electrodes are provided in the discharge tube to produce discharge in the gas. They are connected to a high voltage power supply. The tube is hermetically sealed by inclined windows arranged at its two ends. On the axis of the tube, two mirrors arc arranged externally which form the Fabry-Perot optical resonator. The distance between the mirrors is adjusted to be m λ/2 such that the resonator supports standing wave pattern.

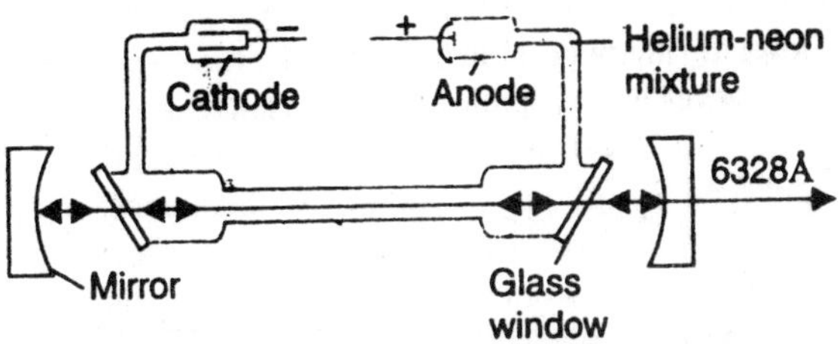

Fig. 18

Working : Helium-Neon laser employs a four-level pumping scheme. The energy levels of helium and neon are. When the power is switched on, a high voltage of about 10 kV is applied across the gas. It is sufficient to ionize the gas. The electrons and ions produced in the process of discharge are accelerated towards the anode and cathode respectively. The energetic electrons excite helium atoms through collisions. One of the excited levels of helium F_3 (2s) is at 20.61 eV above the ground level. It is a metastable level and the excited helium atom cannot return to the ground level through spontaneous emission. However, it can return to the ground level by transferring its excess energy to a neon atom through collision. Such an energy transfer can take place when the two colliding atoms have identical energy levels. Such an energy transfer is known as *resonant energy transfer*.

One of the excited levels of neon E_6 (5s) is at 20.66eV, which is nearly at the same level as F_3 of helium atom. Therefore, resonant transfer of energy can occur between the excited helium atom and ground level neon atom. The kinetic energy of helium atoms provides the additional 0.05 eV required for excitation of the neon atoms. Helium atoms drop to the ground state after exciting neon atoms. This is the pumping mechanism in He-Ne laser. The role of helium atoms is to excite neon atoms and to cause population inversion. The probability of energy transfer from helium atoms to neon atoms is more, as there are 10 helium atoms per 1 neon atom in the gas mixture. The probability of reverse transfer of energy from neon to helium atom is negligible.

The upper state of neon atom E_6 is a metastable state. Therefore, neon atoms accumulate in this upper state. The E_3 (3p) is sparsely populated at ordinary temperatures, and a state of population inversion is readily established between E_6 and E_3 levels. Random photons emitted spontaneously prompt stimulated emission and lasing occurs. The transition $E_6 \rightarrow E_3$ generates a laser beam of red colour of wavelength 6328Å. Other possible transitions produce 3.39 μm and 1.15 μm laser beams respectively. From the level E_3 the neon atoms drop to E_2 (3s) level spontaneously. E_2 level is however a metastable state. Consequently, neon atoms tend to accumulate at E_2 level. It is necessary that these atoms are brought to the ground state E_1(2p) quickly; otherwise the number of atoms at the ground state will go on diminishing and the laser ceases to function. The only way of bringing the atoms to the ground state is through collisions.

If the discharge tube is made narrow, the probability of atomic collisions with the tube walls increases. Because of frequent collisions with the walls, the neon atoms rapidly drop to the ground level and will be available for excitation once again. If the diameter of the discharge tube is increased, the probability of collisions of atoms with the walls decreases and the neon atoms tend to accumulate at energy level E_2. In due course of time, the atoms are no more available at the ground level for further excitation. Therefore, the laser ceases to operate.

He-Ne laser operates in cw mode and is widely used in laboratories as a monochromatic source. It is also widely used in laser printing, bar code reading, etc.

Carbon Dioxide Laser

The carbon gas laser is a very useful and efficient laser. It is a four-level molecular laser and operates at 10.6 μm in far IR region.

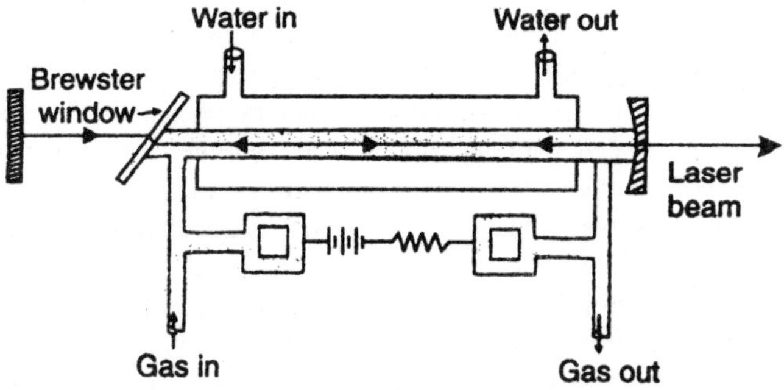

Fig. 19 : Schematic of a carbon dioxide laser.

Construction : The schematic of typical CO_2 laser is shown in Fig. 19. It is basically a discharge tube having a bore of cross section of about 1.5 mm^2 and a length of about 260 mm. The discharge tube is filled with a mixture of carbon dioxide, nitrogen and helium gases in 1:4:5 proportions respectively. Other additives such as water vapour are also added. The active centres are CO_2 molecules lasing on the transitions between the vibrational levels of the electronic ground state.

Energy Levels of CO_2 Molecule

The electron energy levels of an isolated atom are discrete and narrow. However, in case of molecules the energy spectrum is complicated due to many additional features. Each electron energy level is associated with nearly equally spaced vibrational levels and each vibrational level in turn has a number of rotational levels.

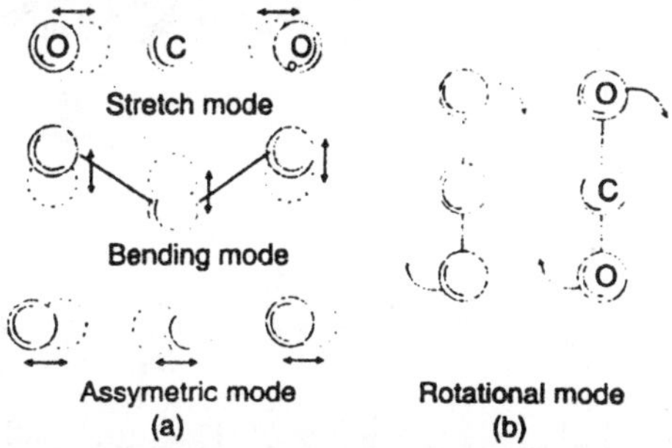

Fig. 20 : Vibrational modes of a CO_2 molecule.

CO_2 molecule is a linear molecule consisting of a central carbon atom with two oxygen atoms attached one on either side. It undergoes three independent vibrational-oscillations known as the *vibrational modes*.

These vibrational degrees of freedom are quantized. At any one time, a CO_2 molecule can vibrate in a linear combination of three fundamental modes. The energy states of the molecule are then represented by three quantum numbers (m n q). These numbers represent the amount of energy associated with each mode. For example, the number (020) indicates that the molecule in this energy state is in the pure bending mode with two units of energy. Each vibrational state is associated with rotational states corresponding to the rotation of CO_2 molecule about its centre of mass. The separations between vibrational-rotational states are much smaller on the energy scale compared to the separations between electron energy levels. The nitrogen molecule N_2 is also characterized by similar vibrational levels. Fig. 20 shows the vibrational modes and rotations of CO_2 molecule.

Working : Fig. 21 shows the lowest vibrational levels of the ground electron energy state of CO_2 molecule and an N_2 molecule. The excited state of an N_2 molecule is metastable and it is identical in energy to (001) vibrational level of CO_2 molecule, indicated as E_5 in Fig. 3.9. As current passes through the mixture of gases, the N_3 molecules get excited to the metastable state. The excited N_2 molecules cannot spontaneously lose their energy and consequently, the number of N_2 molecules at the level keeps on increasing.

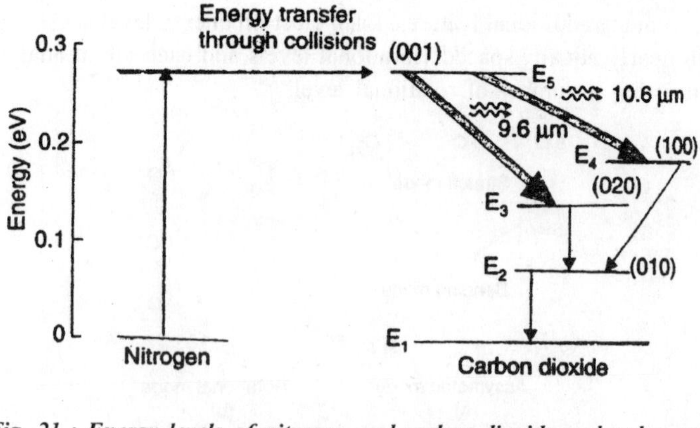

Fig. 21 : Energy levels of nitrogen and carbon dioxide molecules and transitions between the levels.

The N_2 molecules return to ground state through inelastic collisions with ground state CO_2 molecules. Thereby the CO_2 molecules are excited to E_5 level. Some of the CO_2 molecules are also excited to the upper level E_5 through collisions with electrons. The excitation of CO_2 molecules through collisions with excited N_2 molecules is similar to that of helium atoms by neon atoms in helium-neon laser. The E_5 level is the upper lasing level while the (020) and (100) states marked as E_3 and E_4 levels act as the lower lasing levels. As the population of CO_2 molecules builds up at E_5 levels, population inversion is achieved between E_5 level and the levels at E_4 and E_3, The laser transition between $E_5 \rightarrow E_4$ levels produces far IR radiation at the wavelength 10.6 μm (1,06,000Å).

The lasing transition between $E_5 \rightarrow E_3$ levels produces far IR radiation at 9.6 μm (96,000Å) wavelength. E_3 and E levels are also metastable states and the CO_2 molecules at these levels fall to the lower level E_2 through inelastic collisions with normal (unexcited) CO_2 molecules. This process leads to accumulation of population at E_2 level. And also, as the gaseous mixture heats up, the E_2 level, which is close to the ground state, E_1, tends to be populated through thermal excitations. Thus, the de-excitation of CO_2 molecules at the lower lasing level poses a problem and inhibits the laser action. The presence of helium along with CO_2 helps to decrease the population density at E_2 level. It de-exeites CO_2 molecules through inelastic collisions and aids cooling the gaseous mixture through heat conduction. The CO_2 laser operates in CW mode and is capable of generating high powers of the order of several kilowatts at a relatively high efficiency of about 40%. Therefore, it is the most widely used laser. Its applications include use in communications, weaponry and laser fusion.

KINDS OF THE ATOM LASERS

There are several ways in which we can classify lasers into different types. We prefer here to classify the lasers on the basis of the material used as active medium. Accordingly, they are broadly divided into four categories, namely solid state lasers, gas lasers, liquid lasers, and semiconductor diode lasers. Most lasers emit light in the red or IR regions. Lasers work in a continuous mode or in a pulsed mode.

Ruby Laser

Ruby laser belongs to the class of solid state lasers. The term solid state has different meanings in the field of electronics and lasers. A solid state laser is one in which the active centers are fixed in a crystal or glassy material. Solid state lasers are electrically nonconducting. They are also called *doped insulator lasers*.

Historically, the ruby laser was the first laser. It was invented in 1960 by Theodore Mail-nan, U.S.A. The ruby laser rod is in fact a synthetic ruby crystal. AlO_3 crystal, doped with chromium ions at a concentration of about 0.05% by weight. Cr^{3+} ions are the actual active centers and have a set of three energy levels suitable for realizing lasing action whereas aluminium and oxygen atoms are inert. *Construction :* The schematic of a ruby laser is shown in Fig. 22. Ruby rod is taken in the form of a cylindrical rod of about 4 cm in length and 0.5 cm in diameter. Its ends are grounded and polished such that the end faces are exactly parallel and are also perpendicular the axis of the rod. One face is silvered to achieve 100% reflection while the other is silvered to give 10% transmission and 90% reflection. The silvered faces constitute the Fabry-Perot resonator. The laser rod is surrounded by a helical photographic flash lamp filled with xenon. Whenever activated by the power supply the lamp produces flashes of white light.

Working : Ruby laser uses a three-level pumping scheme. The energy levels of Cr^{3+} ions in the crystal lattice are shown in Fig. 22. There are two wide energy bands E_3 and E'_3 and a pair of closely spaced levels at E_2. When the flash lamp is activated, the xenon discharge generates an intense burst of white light lasting for a few milliseconds. The Cr^{3+} ions are excited to the energy bands E_3 and E'_3 by the green and blue components of .white light. The energy levels in these bands have a very small lifetime ($\approx 10^{-9}$s). Hence the excited Cr^{3+} ions rapidly lose some of the energy to the crystal lattice and undergo non-radiative transitions.

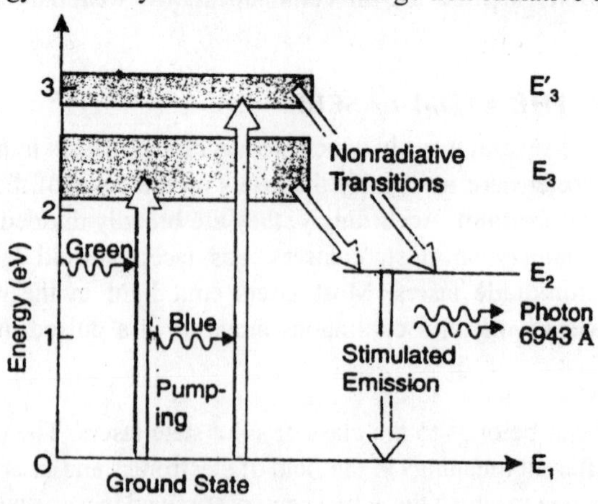

Fig. 22 : Energy levels and transitions in a ruby laser.

They quickly drop to the levels E_2. The pair of levels at E_2 are metastable states having a lifetime of approximately 1000 times more than the lifetime of E_3 level. Therefore, Cr^{3+} ions accumulate at E_2 level. When more than half of the Cr^{3+} ion population accumulates at E_2 level, the state of population inversion is established between E_2 and E_1 levels. A chance photon emitted spontaneously by a Cr^{3+} ion initiates a chain of stimulated emissions by other Cr^{3+} ions in the metastable state. Red photons of wavelength 6943Å travelling along the axis of the ruby rod are repeatedly reflected at the end mirrors and light amplification takes place. A strong intense beam of red light emerges out of the front-end mirror.

Note that the green and blue components of light play the role of pumping agents and are responsible for causing population inversion. The spontaneous photons of λ = 6943Å, corresponding to red colour, act as the input of the oscillator which actually gets amplified. The xenon flash lasts for a few milliseconds. However, the laser does not operate throughout this period. Its output occurs in the form of irregular pulses of microsecond duration. It is because the stimulated transitions occur faster than the rate at which population inversion is maintained in the crystal. Once stimulated transitions commence, the metastable state E_2 gets depopulated very rapidly and at the end of each small pulse, the population at E_2 has fallen blow the threshold value required for sustained emission of light. As a result the lasing ceases and laser becomes inactive. The next pulse appears after the population inversion is once again restored. The process repeats.

Nd: Yag Laser

Nd : YAG laser is one of the most popular types of solid state laser. It is a four-level laser. Yttrium aluminium garnet, $Y_3Al_5O_{12}$, commonly called YAG is an optically isotropic crystal. Some of the Y^{3+} ions in the crystal are replaced by neodymium ions, Nd^{3+}. Doping concentrations are typically of the order of 0.725% by weight. The crystal atoms do not participate in the lasing action but serve as a host lattice in which the active centres, namely Nd^{3+} ions reside.

Construction : The system consists of an elliptically cylindrical reflector housing the laser rod along one of its focus line and a flash lamp along the other focus line. The light leaving one focus of the ellipse will pass through the other focus after reflection from the silvered surface of the reflector. Thus the entire flash lamp radiation gets focused on the laser rod. The YAG crystal rods are typically of 10 cm in length and 12 mm in diameter. The two ends of the laser rod are polished and silvered and constitute the optical resonator.

Working : A simplified energy level diagram for the neodymium ion in YAG crystal. The energy level structure of the free neodymium atom is preserved to a certain extend because of its relatively low concentration. However, the energy levels are split and the structure is complex. It is essentially a four-level system with the terminal laser level E_2 sufficiently far removed from the ground level. The pumping of the Nd^{3+} ions to upper states is done by a krypton arc lamp. The optical pumping with light of wavelength range of 5000 to 8000Å excites the ground state Nd^{3+} ions to the multiple energy levels at E_4. The metastable level E_3 is the upper laser level, while the E_2 forms the lower laser level. The upper laser level E_3 will be rapidly populated, as the excited Nd^{3+} ions quickly make downward transitions from the upper energy bands. The lower laser level E_2 is far above the ground level and hence it cannot be populated by Nd^{3+} ions through thermal transitions from the ground level. Therefore, the population inversion is readily achieved between the E_3 level and E_2 level. The laser emission occurs in infrared (IR) region at a wavelength of about 10,600Å (1.06 μm). As the laser is a four level laser, the population inversion can be maintained in the face of continuous laser emission.

MECHANISM OF LIGHT EMISSION

Radiation comes in a broad range of frequencies grouped into radiowaves, microwaves, infrared rays, visible rays, ultraviolet rays, x-rays and γ-rays. They are all essentially electromagnetic waves and therefore, the mechanism of their emission should be the same. In all cases electromagnetic waves originate in the oscillations of electric charge. The simplest device, which produces electromagnetic waves, is an *oscillating dipole.* A dipole consists of two opposite charges. When these opposite charges vibrate to and fro along a straight line, they constitute an oscillating dipole. In terms of the simple Bohr's theory light emission from atoms is explained as due to electron's transition from an excited state to the ground state. According to quantum-mechanical considerations, the electric dipole moment of the atom is the source of light and one can calculate the probability of electron making a transition. We may visualize that during the act of light emission the orbital electron makes its downward transition through a gradually damped oscillatory motion occurring at the specific resonance frequency. In this chapter, we briefly describe some of the important developments that led to the understanding of atomic structure and mechanism of light emission.

SOLVED EXAMPLES

Example 1:

A laser source of λ = 6000Å, coherence width 4 mm and power 10mW shines on a surface 100m away. Deduce the illumination. Compare it with that due to a focussed beam from a torch of filament diameter 1mm, power 10W, lens focal length 10 cm and aperture 4 cm dia.

Solution:

For the laser all 100mW power goes in a cone of semi angle $\theta = \lambda/a = 6 \times 10^{-5}/0.4 = 15 \times 10^{-5}$ rad. Hence a real spread at 10^4 cm distance (besides the initial area $\pi \times 0.2^2$ cm^2) is –

$$A = D^2\pi\,\lambda\theta^2 = 10^8 \times 2.14 \times (15 \times 10^{-5})^2 = 7\text{cm}^2$$

$$\text{Illumination} = \text{Power/Area} = 10\text{mW}/7\text{ cm}^2 = 1.4\text{ mW/cm}^2$$

For the torch the 10W power spreads isotropically, and the lens collects a fraction π (2 cm)^2 ÷ 4π (10 cm)^2 = 1/100. The light spreads over a cone of semiangle determined by the filament size ÷ focal length of the lens = 0.1/10 = 1/100 rad. Hence a real spread at distance 10^4 cm is

$$A' = D^2\pi\,\theta^2 = 10^8 \times 3.14 \times (10^{-2})^2 = 3.1 \times 10^4\text{ cm}^2$$

$$\text{Illumination} = \text{Power/Area}$$

$$= \frac{1}{100} \times 10\text{W}/3.1 \times 10^4\text{cm}^2 = 0.003\text{ mW/cm}^2.$$

Example 2:

If the laser beam of Ex. 21.3 is focussed by a lens of focal length 10 cm, deduce (i) the radius, (ii) area, and (iii) power density of the image.

Solution:

Angular spread of the laser beam is – l/d, due to diffraction alone.

Hence $\theta = 6 \times 10^{-5}/0.4 = 1.5 \times 10^{-4}$ rad

$\therefore$ radius of the image = $f\theta = 1.5 \times 10^{-3}$ cm area of the image = $\pi\ (1.5 \times 10^{-3})^2 = 8 \times 10^{-6}\ cm^2$

power density = $\dfrac{10\times10^{-3}\,W}{8\times10^{-6}\,cm^2} = 1.2\ KW/cm^2$.

Example 3:

With certain units for P and E, let the three χ coefficients of the order of 10^{-4}, 10–9 and 10^{-16} respectively. evaluate the relative contributions of the three terms for :

(a) E = 1 unit,

(b) E = 104 units,

(c) E = 107 units.

Solution:

Simple substitution gives for the three cases:

(a) 1 : 10^{-5} : 10^{-12}

(b) 1 : 10^{-1} : 10^{-4}

(c) 1 : 10^{+2} : 10^{+2}

Note: $\chi^{(1)}$, $\chi^{(2)}$, $\chi^{(3)}$, are dimensionally different. Hence their relative numerical values have no meaning by themselves; they depend on the units for E and P. Thus, if the unit for E were taken 10^6 – fold larger (numerical E values 10^{-6} fold), then in the same example $\chi^{(1)}$, $\chi^{(2)}$, $\chi^{(3)}$, would be 10^{+2}, 10^{+3} 10^{+5} respectively.

Example 4:

Deduce $\Delta\nu$ values per angstrom difference of wavelength at

λ = 2500Å, and at λ = 5000Å.

Solution:

$$\bar{\nu} = 1/\lambda \Rightarrow \Delta\bar{\nu} = -\Delta\lambda/\lambda^2$$

For $\Delta\lambda = 1 \times 10^{-8}$ cm at $\lambda = 2500 \times 10^{-8}$ cm

$$\Delta\bar{\nu} = (1 \times 10^{-8})/(2.5 \times 10^{-5})^2 = -16\ cm^{-1}$$

For $\Delta\lambda = \lambda \times 10^{-8}$ cm at $\lambda = 5000 \times 10^{-8}$ cm

$$\Delta\bar{\nu} = -(1 \times 10^{-8})/(5 \times 10^{-5})^2 = -4\ cm^{-1}$$

(The minus sign only means that for greater l the $\bar{\nu}$ value is smaller).

Example 5:

Excited by 5460.7Å radiation a sample gives a Raman line at 5542.6Å. Compute:

(i) the Raman frequency,

(ii) position of the corresponding anti-stokes line.

Solution:

$$v_o = \frac{10^8}{5460.7} = 18312\ \text{cm}^{-1}$$

$$v_R = \frac{10^8}{5542.6} = 18042\ \text{cm}^{-1}$$

$\therefore$ Raman frequency $\Delta v_R = v_o - v_R = 270\ \text{cm}^{-1}$

anti-stokes frequency $v'R = v_o + \Delta v_R$

$$= 18312 + 270 = 18582\ \text{cm}^{-1}$$

$\therefore$ anti-stokes wavelength $\lambda' = \dfrac{10^8}{v'R} = 5381.5\text{Å}.$

Example 6:

At what temperature are the rates of spontaneous and stimulated emission equal? Assume λ = 5000 Å.

Solution:

If the ratio of spontaneous and stimulated emission are equal, the

$$R_1 = \left[\frac{1}{e^{hv/kT} - 1}\right] = 1 \quad \text{or} \quad e^{hv/kT} = 2$$

As λ = 5000 Å, $v = c/\lambda = 6 \times 10^{14}$Hz and

$$\frac{hv}{kT} = \frac{6.626\times10^{-34}\,\text{Js}(6\times19^{14}/\text{s})}{(1.38\times10^{-23}\,\text{J/K})\text{T}} = \frac{28.8\times10^3}{\text{T}}\text{K}$$

$$e^{hv/kT} = \exp\left[\frac{28.8\times10^3}{\text{T}}\text{K}\right] = 2$$

or $$\frac{28.8\times10^3}{\text{T}}\text{K} = \text{In } 2 = 0.693$$

$\therefore$ $$\text{T} = \frac{28.8\times10^3}{0.693}\text{K} = 41{,}558\ \text{K}.$$

Example 7:

At what frequency should an electron revolve around a proton if orbit radius is 0.53Å? Estimate the wavelength of radiations emitted on the classical theory.

Solution:

Equating centripetal force with the Coulomb force,

$$\frac{mv^2}{r} = \frac{e^2}{4\pi \epsilon_0 r^2} = \Rightarrow v = \left(\frac{e}{4\pi \epsilon_0 nr}\right)^{1/2}$$

Frequency is given by $v/2\pi r$. Hence,

$$f = v/2\pi r = [e/2\pi]\ (4\pi\epsilon_0 m)^{-12} r^{-3/2}$$

with $r = 0.53 \times 10^{-8}$ cm, substitution gives

$$f = 6.6 \times 10^{15}\ s^{-1}$$

On the classical theory this is also the frequency of radiation emitted. Hence

$$\lambda = \frac{c}{f} = \frac{3.0 \times 10^{10}\ cms^{-1}}{6.6 \times 10^{15}\ s^{-1}} = 4.5 \times 10^{-6}\ cm\ (450\ A^o).$$

Example 8:

The angular diameter of a star is negligibly small compared to the diffraction spread λ/a due to viewing with an instrument of aperture diameter a. If the effective brightness of the star is 10^{-4} times that of the background sky in viewing with the naked eye, what will the relation be in viewing through a telescope of 200 times the diameter of the pupil of the eye?

Solution:

We will use subscripts s and o for the star and background sky, and primes' and" for viewing with the eye and the telescope respectively. We have the general relations.

$$I = BA = B'A' = B''A''$$

and $A = r^2\Omega$

where B is the brightness, A is the transverse area, r is the distance, and W the solid angle. For the star, Ω arises due primarily to diffraction, so that

$$\Omega'_s = \pi\,(\lambda/a')^2;\ \Omega''_s = \pi\,(\lambda/a')^2$$

Substitutions lead to

$$B''_s = B'_s\,(a''/a')^2$$

In the case of the sky, the intrinsic Ω_o far exceeds the contributions due to diffraction, and hence $\Omega''_o = \Omega'_o = \Omega_o$. That means

$$B''_o = B'_o$$

Hence $\quad (B''_s B''_o) = (B'_s/B'_o)\,(a''/a')^2 = (10^{-4})\,(200)^2 = 4$

Thus, in the telescope the star brightness is 4 times the sky brightness.

(*Note:* It is incorrect to take the effect as due to larger light gathering by a telescope lens. That effect is common for the star and the background light. The effect is because of smaller diffraction spread in star image as a increases).

Example 9:

A plane diffraction grating with 5000 lines/cm is used in Littrow mounting with angle of incidence 20° and focal length of collimator/camera lens is 50 cm. Deduce for the second order spectrum :

(i) The wavelength falling at the centre of the field of view,

(ii) Linear dispersion in the spectrum (in Angstrom/mm).

Solution:

(i) grating element e = 1/5000 cm = 2×10^{-4} cm. In the Littrow mount $\theta = i = 20°$ (given). Hence, for second order,

$$2\lambda = e\,(\sin i + \sin\theta) = 2 \times 10^{-4}\,(0.34 + 0.34)$$

$$\therefore\quad \lambda = 6.8 \times 10^{-5}\text{ cm} = 6800\text{Å}.$$

(ii) Angular dispersion $d\theta/d\lambda$ for Littrow mount is given by

$$e\cos\theta.d\theta = n d\lambda$$

$\therefore$ Linear dispersion dx/dl is given by

$$\frac{dx}{d\lambda} = f.\frac{d\theta}{d\lambda} = f\,\frac{n}{e\cos\theta}$$

$$= \frac{50\times2}{2\times10^{-4}\times0.94}\,x10^{-8}\ \text{cm/angstrom}$$

$$= 5.3 \times 10^{-2}\ \text{mm/A}$$

$$\therefore\quad d\lambda/dx = 19\ \text{A/mm}.$$

Example 10:

A 100 W lamp may be treated as emitting radiations at different wavelengths as listed below:

$\lambda/1000\ A$ = 4.8	5.2	5.6	6.0	6.4	6.8	7.2
$P\lambda/Watt$ = 0.4	0.8	1.2	1.6	2.0	2.8	3.0

The rest of the power is spent otherwise. Using the sensitivity curve of the human eye, and given μ_{max} = 683 lm/W, deduce the μ of the lamp.

Solution:

We list the corresponding Kλ values and hence PλKλ

K_λ = 0.20	0.71	0.95	0.60	0.16	0.04	0.00
$P_\lambda K_\lambda$ = 0.0.8	0.57	1.14	0.96	0.32	0.11	0.00

That gives $\Sigma P_\lambda K_\lambda$ = 3.2W. Hence

$$\mu = \frac{3.2\text{W}\times 6.83\text{lm / W}}{100\text{W}} = 22 \text{ lm/W.}$$

Example 11:

A typical coiled-coil filament lamp of 100 cm intensity has a filament of apparent length 2.5 cm and diameter 0.30 cm. Deduce the total flux emission, luminosity and brightness, assuming isotropic character of emission.

Solution:

Total flux = $Id\Omega$ = $I.4\pi$ = 400π lm

Area of cross-section = $2.5 \times 0.30 \times 10^{-4}\text{m}^2 = 7.5 \times 10^{-5}\text{m}^2$

Luminosity ($L_{average}$)

$$= \frac{\phi}{S} = \frac{400\pi \text{ lm}}{7.5\times 10^{-5}\text{ m}^2} = 5.3\pi \times 10^6 \text{ lm/m}^2$$

Brightness ($B_{average}$)

$$= \frac{I}{S} = \frac{400\text{cd}}{7.5\times 10^{-5}} = 5.3 \times 10^6 \text{ cd/m}^2$$

(*Note:* L here is –9 times that for the surface defining the candela. This is not because of higher temperature – T is of same order – but is be cause of coiled coil. The actual coil is much longer than that given as apparent length here).

Example 12:

Deduce the wavelength of the first two Balmer lines in He+, given $R = 1.097 \times 10^5\ cm^{-1}$.

Solution:

For He^+ we have Z = 2. Hence the Balmer series is

$$\bar{v} = 1.097 \times 10^5 \times 4 \left(\frac{1}{2^2} - \frac{1}{n^2}\right) cm^{-1};\ n = 3, 4, ...$$

Substituting n = 3 and 4 in turns and taking the reciprocal gives:

$$\lambda = 1.64 \times 10^{-5} \text{ cm}; 1.21 \times 10^{-5} \text{ cm.}$$

Example 13:

The length of a laser tube is 150 mm and the gain factor of the laser material is 0.0005/cm. If one of the cavity mirrors reflects 100% light that is incident on it, what is the required reflectance of the other cavity mirror?

Solution:

$$\gamma_{th} = \frac{1}{2L} \text{In} \frac{1}{r_1 r_2}$$

$$\therefore \quad r_2 = \frac{1}{r_1 e^{2L\gamma}} = \frac{1}{1 \times e^{2 \times 15 \times 0.0005}} = 0.985$$

It means that the second mirror should have a reflectance of 98.5%.

Example 14:

Deduce the de Broglie wavelength of :

(i) neutrons of kinetic energy 1 ev,

(ii) thermal neutrons.

Solution:

(i) Since $p = \sqrt{2mE}$, we get (using 1 ev = 1.6×10^{-19} joule)

$$\lambda = \frac{h}{\sqrt{2mE}} = \frac{6.6 \times 10^{-34}}{(2 \times 1.67 \times 10^{-27} \times 1.6 \times 10^{-19})^{1/2}}$$

$$= 2 \times 10^{-11} \text{ m}$$

(ii) For *thermal neutrons, energy* E = kT. We get

$$E = 1.38 \times 10^{-23} \times 300 = 4 \times 10^{-21} \text{ joule}$$

$$\lambda = \frac{6.6\times10^{-34}}{(2\times1.67\times10^{-27}\times4\times10^{-21})^{1/2}} = 2 \times 10^{-10} \text{ m.}$$

Example 15:

Make an order of magnitude calculation of the de Broglie wavelengths of a 100-volt electron (i.e., an electron accelerated through 100 volt potential difference).

Solution:

The momentum p is given by the relation $p^2 = 2mE$,

$$p = (2 \times 9.1 \times 10^{-31} \times 1.6 \times 10^{-19} \times 100)^{1/2}$$

$$= 5 \times 10^{-24} \text{ kg ms}^{-1}$$

Hence, de Broglie wavelength is

$$\lambda = \frac{h}{p} = \frac{6.6\times10^{-34}}{5\times10^{-24}} = 1.3 \times 10^{-10} \text{ m.}$$

Example 16:

The half-width of the gain profile of a He-Ne laser material is 2×10^{-3} nm. If the length of the cavity is 30 cm, how many longitudinal modes can be excited? The emission wavelength of He-Ne laser is 6328 Å.

Solution:

The separation between successive longitudinal modes is given by

$$\Delta\lambda = \frac{\lambda^2}{2L} = \frac{(6328\times10^{-10}\text{m})^2}{2(30\times10^{-2}\text{m})}$$

$$= 0.66 \times 10^{-3} \text{ nm}$$

$$\text{Number of modes } N = \frac{\delta\lambda}{\Delta\lambda}$$

$$= \frac{2\times10^{-3}\text{nm}}{0.66\times10^{-3}\text{nm}} = 3.$$

Example 17:

The coherence length for sodium D_2 line is 2.5 cm. Deduce :

(i) the coherence time τ,

(ii) the spectral width of the line,

(iii) the purity factor (Q).

Solution:

We have

$$t = \frac{L}{c} = \frac{2.5\text{cm}}{3.0\times10^{10}\text{cm/s}} = 0.8 \times 10^{-10}\text{s}$$

From Eq. (21.3b), taking $\lambda = 6 \times 10^{-5}$ cm

$$\Delta\lambda = \frac{\lambda^2}{L} = \frac{36\times10^{-10}\text{cm}^2}{2.5\text{cm}} = 14 \times 10^{-10} \text{ cm} = 0.14 \text{ A}$$

$$\text{We have, } Q = \frac{\lambda}{\Delta\lambda} = \frac{6\times10^{-5}\text{cm}}{14\times10^{-10}\text{cm}} = 0.4 \times 10^5.$$

Example 18:

With a He-Ne laser, Michelson interferometer fringes remained clearly visible when the path difference was increased upto 8 m. Deduce the lower limits for :

(i) the coherence length,

(ii) coherence time,

(iii) spectral half width, and

(iv) Q of the line, (Given $l = 11.5 \times 10^{-5}$ cm).

Solution:

The experimental fact of observing interference for path difference upto 8 m directly gives–coherence length L > 8m

$$\text{Then give–coherence time } \tau = \frac{L}{c} > 2.7 \times 10^{-8} \text{ sec}$$

$$\text{Spectral half-width } \Delta\lambda = \frac{\lambda^2}{L} = 1.6 \times 10^{-3}\text{Å}$$

$$Q = \frac{L}{\lambda} = 7 \times 10$$

Example 19:

One sectorial series in an element is given by

$$v_n = \frac{109756}{(2-0.083)^2} - \frac{109756}{(n-0.643)^2}, \; n = 3, 4, 5, \ldots$$

Write down two more spectral series using Ritz principle, and evaluate v for the fist member of each series.

Solution:

We predict one series by changing the integer in the first term denominator to 3. Thus

$$v_n' = \frac{109756}{(3-0.083)^2} - \frac{109756}{(n-0.0643)^2},\ n = 4, 5, 6, ...$$

We predict one more series by using the first running term of the given series as the fixed term, and changing the fixed term to running terms. Then

$$v_n'' = \frac{109756}{(3-0.643)^2} - \frac{109756}{(n-0.083)^2}\ n = 3, 4, 5, ...$$

For the new series thus predicted the v values of the first term may be now calculated

$$v' = 3.17 \times 10^3\ cm^{-1};$$

$$v'' = 6.85 \times 10^3\ cm^{-1}.$$

Example 20:

About 0.1% of electrical energy supplied to a laboratory mercury vapour lamp of 80 watt is converted into UV light of wavelength 2500Å. Calculate the number of UV photons emitted per second by the lamp.

Solution:

Number of UV photons emitted per second

$$= \frac{\text{Energy conveted int o UV light}}{\text{Energy carried by one UV photon}}$$

$$= \frac{80J/s \times (0.1/100) \times (6.24 \times 10^{18} eV/J)}{12400eV/2500}$$

$$= 10^{17}\ \text{photons/s}.$$

Example 21:

Deduce the average energy of Planck's oscillator corresponding to v values such that hv/kT = 0.1, 1.0, 3.0, 10.0.

Solution:

We have

$$\bar{\varepsilon} = kT \frac{x}{e^x - 1}, \text{ where } x = \frac{h\nu}{kT}$$

For the four given vales of x, tables show

$\varepsilon^{x-1} = 0.105,\ 1.718,\ 19.1, \text{ and } 2.2 \times 10^4$

Hence the value of ε is given by

$$\bar{\varepsilon} = kT\left(\frac{.1}{.105}, \frac{1}{1.718}, \frac{3}{19.1}, \frac{10}{2.2\times10^4}\right)$$

$$= kT\ (1, 0.58,\ 0.16,\ 0.00045)$$

Thus for hν, < < kT we have ε = kT while for hν > > kT, ε tends to zero.

Example 22:

A surface is placed 20 cm from a 50W source of λ = 6x 10⁻⁵ cm. Estimate (i) the average time interval between the arrival of two photons on the same atom (sectional area – $10^{-16}cm^2$, and (ii) the mean time lag between irradiation and ejection of the first electron from the surface (area 2 cm^2, efficiency 40%).

Solution:

Energy flux = $50/4\pi\ (0.20)^2 = 100\ Jm^{-2}s^{-1}$

Photon energy $= hc/\lambda = 6.6 \times 10^{-34} \times 3 \times 10^{+8}/6 \times 10^{-7}$

$= 3 \times 10^{-19}$ J/photon

∴ Photon flux at 20 cm = $100/(3 \times 10^{-19})$

$= 3 \times 10^{20}$ photon $m^{-2}s^{-1}$

Now, (i) photon flux on a given atom = 3 photon s^{-1}

λ mean $\Delta t = 0.3s$

(ii) ejection rate from 2 cm^2 area = $3 \times 10^{20} \times 2 \times 10^{-4} \times .4$

$= 2.4 \times 10^{16}$ electrons^{-1}

∴ mean $\Delta t = 4 \times 10^{-15}s$

(*Note*: One conclusion is that the chances of two photons acting together in ejecting an electron are negligibly small. Another is that the time lag between irradiation and start of emission is too small to be observable.)

Example 23:

X-rays of 0.5Å are scattered by free electrons in block of carbon through 90°. Find the velocity of recoil electrons.

Solution:

K.E. of the recoil electrons = $h\nu_i - h\nu_f$

$$= 12400\left(\frac{1}{\lambda_i} - \frac{1}{\lambda_f}\right)\text{eV } (1.602 \times 10^{-19} \text{ J/e})$$

$$\lambda_f = \lambda_i + \frac{h}{m_0 c}(1-\cos\theta)$$

$$= (0.5 + 0.02426)\text{Å} = 0.5243\text{Å}$$

$$\therefore \quad \text{K.E.} = 12400\left(\frac{1}{0.5} - \frac{1}{0.5243}\right)(1.602 \times 10^{-19})\text{ J}$$

$$\therefore \quad u = \left[\frac{2(\text{K.E.})}{m}\right]^{1/2} = \left[\frac{2(1.84\times10^{-16}\text{J})}{9.11\times10^{-31}\text{Kg.}}\right]^{1/2}$$

$$= \left\{4.04\times10^{14}\frac{\text{kg.m}^2/\text{s}^2}{\text{kg}}\right\}^{1/2}$$

$$\therefore \quad u = 2 \times 10^7 \text{ m/s.}$$

Example 24:

Compute the minimum uncertainty in the location of a mass of 2.0gm moving with a speed of 1.5 m/s and the minimum uncertainty in the location of an electron moving with a speed of 0.5×10^8 m/s. Given that the uncertainty in the momentum p for both $\Delta p = 10^{-3}p$.

Solution:

$$p = mu = 2 \times 10^{-3}\text{ kg} \times 1.5\text{ m/s} = 3 \times 10^{-3}\text{ kg m/s.}$$

$$\therefore \quad \Delta p = 10^{-3}\text{ p} = 3 \times 10^{-6}\text{ kg m/s}$$

$$\therefore \quad \Delta x = \frac{h}{2\pi\Delta p} = \frac{6.63\times10^{-34}\text{Js}}{2\pi\times3\times10^{-6}\text{kg.m/s}}$$

$$= 3.5 \times 10^{-19}\text{Å}$$

For an electron, p = mu

$$= (9.11 \times 10^{-31}\text{ kg})(0.5 \times 10^8\text{ m/s})$$

$$= 4.55 \times 10^{-25}\text{ kg. m/s}$$

$\therefore \quad \Delta p = 10^{-3}\ p = 4.55 \times 10^{-26}$ kg. m/s

$$\therefore \quad \Delta x = \frac{h}{2\pi\Delta p} = \frac{6.63\times10^{-34}\,\text{Js}}{2\pi\times4.55\times10^{-26}\,\text{kg.m/s}} = 23\text{Å}.$$

Example 25:

With Franck-Hertz type of experiment on sodium vapour the first spectral line to appear is the D-line, λ = 5.89x 10^{-5} cm. Deduce the first excitation potential of sodium.

Solution:

If excitation potential is V volt, excitation energy is eV joule where = 1.6×10^{-19} coul. Equating this with hv,

$$1.6 \times 10^{-19}V = hv = \frac{hc}{\lambda} = \frac{6.62\times10^{-34}\times3\times10^{8}}{5.89\times10^{-7}}$$

$$V = \frac{6.62\times3}{5.89\times1.6} = 2.1 \text{ volt.}$$

Example 26:

For a given metal surface the limiting wavelength for photoelectric emission if 5800Å. Deduce the work function, maximum kinetic energy of electrons ejected by light of λ = 4500A.

Solution:

$$\text{Photon energy } hv = \frac{6.6\times10^{-34}\times3\times10^{10}}{4500\times10^{-8}} = 4.40 \times 10^{-19}\text{J}$$

$$\text{Work function} = hv_o = \frac{6.6\times10^{-34}\times3\times10^{10}}{5800\times10^{-8}} = 3.41 \times 10^{-19}\text{J}$$

The difference gives Ek = 0.99×10^{-19}J

Example 27:

Light of wavelength 5461A and intensity 0.20×10^{-5} watt/cm^2 falls on a photoelectric cell of cathode area 2.5 cm^2. The efficiency is 15%. Compute the saturation photocurrent.

Solution:

Photons received *per second*

$$n = \frac{0.20\times10^{-5}\times2.5}{6.6\times10^{-34}\times3\times10^{8}\times1/(5461\times10^{-10})} = 1.37 \times 10^{13}\text{s}^{-1}$$

Electrons ejected per second are 15% of this; and each electron carries a charge 1.6×10^{-19} coulomb. Hence the saturation current is

$$i = 1.37 \times 10^{13} \times 0.15 \times 1.6 \times 10^{-19} \text{ amp}$$

$$= 0.33 \text{ microamp.}$$

Example 28:

In an experiment on Compton scattering the wavelength of incident radiation is 1.872 A. Calculate the wavelength of scattered radiation at θ = 30°. Also calculate the velocity of the corresponding recoil electron.

Solution:

By simple substitution for θ = 30°

$$\Delta\lambda = .0242\ (1\text{-}8660) = .00324\text{A}°;$$

$$\lambda' = 1.872 + .003 = 1.875.$$

The energy difference of the incident and scattered photons gives the KE of the ejected electron. Hence

$$\frac{1}{2}mv^2 = h\ (v - v') = \frac{hc}{\lambda^2}(\Delta\lambda)$$

$$= \frac{6.6\times10^{-27}\times3\times10^{10}}{(1.872\times10^{-8})^2} \times (.00324 \times 10^{-8}) \text{ erg}$$

Substituting for m and solving we get

$$v = 2 \times 10^8 \text{ cm/sec.}$$

Now, from conservation of momentum, remembering that √/v = 1.

$$mv \cos\phi = \frac{hv}{c} - \frac{hv'}{c} \times .866$$

$$= \frac{hv'}{c} \times 0.134$$

$$mv \sin\phi = \frac{hv'}{c} \text{x } 0.500$$

$$\therefore \quad \tan\phi = \frac{0.5}{0.134} = 3.62 \Rightarrow \phi = 75°.$$

Example 29:

With α-particles from polonium (v = 1.6 × 10⁹ cm/sec) calculate the nearest approach r_o with silver nuclei (Z = 4.7).

Solution:

Nearest approach is for head-on collision, and all kinetic energy then becomes potential energy.

$$\frac{1}{2}m_\alpha v^2 = \frac{2e.Ze}{4\pi\varepsilon_o r_0}$$

$$\therefore \quad r_o = \frac{1}{4\pi\varepsilon_o}\cdot\frac{4Ze^2}{m\alpha v^2} = (9 \times 10^9)\ \frac{4\times 47\times(1.6\times 10^{-19})^2}{(4\times 1.67\times 10^{-27}(1.6\times 10^7)^2} m$$

$$= 2.5 \times 10^{-13} \text{ cm.}$$

Example 30:

In a long chain organic molecule of length 5Å electrons may bet created as free to move along the length. Deduce the zero point energy, the energy gap between the first two energy states of the electron, and also the wavelength of absorption line arising from this transition.

Solution:

$$E_1 = \frac{h^2}{8ma^2} = \frac{(6.6\times 10^{-34})^2}{8\times 9.1\times 10^{-31}\times(5\times 10^{-10})^2} = 2.4 \times 10^{-19} J$$

This is the zero-point energy. Its value *per mole* comes to

$$NE_1 = 1.5 \times 15^5 \text{ J/gm mole.}$$

$$E_2 = 4.\ E_1 = 9.6 \times 10^{-19} \text{ J}$$

$$\therefore \quad \text{Energy gap } \Delta E = 7.2 \times 10^{-19} \text{ J}$$

$$= \frac{7.2\times 10^{-19}}{1.6\times 10^{-19}} eV = 4.5 \text{ eV}$$

Absorption wavelength l is given by

$$\Delta E = hv = hc/\lambda$$

$$\lambda = \frac{hc}{\Delta E} = \frac{6.6\times 10^{-34}\times 3.0\times 10^8}{7.2\times 10^{-19}} m = 2.8 \times 10^4 Å$$

[Note results have been checked quantitatively in the absorption spectra of several organic molecules].

Example 31:

Harmonic 2ω is to generated out of a laser beam ω = 4 × 10^{14} Hz. If the medium has refractive index 1.528 at ω and 1.542 at 2ω, deduce *the momentum mismatch due to colour dispersion. Also estimate the length of the medium for which the 2ω beam will remain coherent if the mismatch is not compensated.*

Solution:

We have $$k = \frac{2\pi}{\lambda} = \frac{2\pi v}{c'} = \frac{\omega}{c}n$$

where c is speed of light in the medium and n is the appropriate refractive index. Substitution leads to

$$\Delta k = \frac{8\times10^{14}\times1.542}{3\times10^{8}} - 2\,\frac{4\times10^{14}\times1.528}{3\times10^{8}}$$

$$= 4 \times 10^{4}\text{m}^{-1}$$

$$\Rightarrow \qquad L = (\Delta k)^{-1} = 2 \times 10^{-5}\text{m}.$$

Example 32:

Find the ratio of populations of the two states in a He-Ne laser that produces light of wavelength 6328 Å At 27°C.

Solution:

The ratio of population is given by $\frac{N_2}{N_1} = e^{-(E_2-E_1)/KT}$

$$E_2 - E_1 = \frac{12400}{6328}\text{eV}$$

$$\therefore \quad \frac{N_2}{N_1} = \exp\left[\frac{-1.96\,\text{eV}}{(8.61\times10^{-5}\,\text{eV})(300\text{K})}\right] = e^{-75.88} = 1.1 \times 10^{-33}.$$

Example 33:

Deduce the Duane and Hunt limit for an X-ray tube working at 40 kV.

Solution:

We have

$$v_{max} = \frac{40000\times1.6\times10^{-19}}{6.62\times10^{-34}}\text{Hz} = 0.97 \times 10^{19}\text{Hz}$$

$$\lambda_{min} = \frac{c}{v_{max}} = 0.31 \times 10^{-8}\text{ cm}.$$

Example 34:

A lamp of I = 400 cd hangs 1.5 m above the centre of a table of dimensions 2.0m × 2.0m. Deduce the illumination at a diagonal end of the table. If a paper placed there gives diffuse scattering with 60% efficiency, deduce the luminance of the paper and also the brightness.

Solution:

For the diagonal end,

$$r^2 = [(1.5)^2 + (1^2 + 1^2)] = 4.25\ m^2;$$

$$\cos\theta = \frac{1.5}{r}$$

$$\text{Illuminance } E = \frac{I\cos\theta}{r^2} = \frac{400 \times 1.5}{(4.25)^{3/2}} = 70lx$$

$$\text{Luminance } L = 0.60 \times E = 42\ lm/m^2$$

$$\text{Brightness } B = L/\pi = 14\ cd/m^2.$$

Example 35:

A source of light has intensity depending on q according to $I = I_o$ cos (θ/2). Deduce the total flux emitted by it, and the averaged intensity as a point source.

Solution:

The solid angle between θ and θ + dθ is 2π sin θ dθ.

Hence $d\phi = 12\pi \sin\theta\, d\theta = I_0 2\pi \sin\theta \cos 1/2\, \theta\, d\theta$

Integrating from θ = 0 to θ = π (point source)

$$\phi = 2\pi I_o \left[-\frac{4}{3}\cos^3\frac{\theta}{2}\right]_\theta^\pi = \frac{8\pi}{3} I_o$$

$$I_{av} = \frac{\phi}{4\pi} = \frac{2}{3} I_o.$$

Example 36:

A source P and an observer Q are placed at angles 3 0° and 60° respectively from the normal to a surface. The source has intensity 1400 cd in the concerned direction, and its distance from the surface is 2.5m. The surface gives diffuse scattering of the incident light with 80% efficiency. Deduce the brightness of the surface for the viewer.

Solution:

The illuminance is

$$E = \frac{I\cos\theta}{r^2}$$

$$= \frac{1400\times\cos 30}{(2.5)^2}\,\text{lm}/\text{m}^2$$

With 80% efficiency, the luminance must be 0.8 E. Also, for an ideal diffuse scatterer, B is the same in all directions, with the relation $L = \pi B$. That gives

$$B = \frac{L}{\pi} = \frac{0.8E}{\pi}$$

$$= \frac{0.8\times 1400\times\cos 30}{\pi(2.5)^2}\,\text{cd}/\text{m}^2.$$

(Notice that angle 60° irrelevant for an ideal diffuse scatterer).

EXERCISES

1. Describe the similarities and differences between an optical laser and an atom laser.
2. The CO_2 laser is one of the most powerful lasers. The energy difference between the two laser levels is 0.117 eV. Determine the frequency and wavelength of the radiation.
3. Find the ratio of populations of the two states in a He-Ne laser that produces light of wavelength 6328Å at 27°C.
4. Describe any two important applications of atom laser.
5. Explain how atom laser is used in holography?
6. A laser beam can be focused on an area equal to the square of its wavelength. For a He-Ne laser, the wavelength of emitted light is 6328Å. If the laser radiates energy at the rate of 1 mW, find out the intensity of the focused beam.
7. Find the relative populations of the two states in a ruby laser that produces a light beam of wavelength 6943Å at 300K.
8. Explain the four-wave mixing of matter waves.
9. Explain in brief the method of cooling a dilute gas to nanokelvin temperatures.

10. What do you understand by an optical resonant cavity? Explain.
11. Why is the optical resonator required in lasers? Illustrate your answer with neat sketches.
12. Deduce the luminance of the sun, given that its disc subtends an angle 30' arc and the illuminance due to it at earth's surface at mid-day is 10^5 lux. An image of the sun is formed by a lens of diameter 8 cm and focal length 30 cm; deduce the illuminance of the image.
13. In a street lighting system, the poles are 12 m apart and the lamps 5.0m above the road. The lamps are designed to give uniform intensity over a solid angle 1.5π sr and zero outside. If the averaged intensity of the lamps is 600 cd each, deduce :
    (a) the intensity within the 15π sr solid angle cone,
    (b) the illuminance at the bottom of each lamp, and
    (c) the illuminance at the mid-point between the bases of adjacent poles.
14. Explain the principle and working of a He-Ne laser.
15. In helium-neon laser lasing is through neon gas. What is then the role of helium gas?
16. With the help of energy band diagram discuss the working of a semiconductor laser.
17. Explain in brief the characteristics of a laser beam.
18. The He-Ne system is capable of lasing at several different IR wavelengths, the prominent one being 3.3913μm. Determine the energy difference (in eV) between the upper and lower levels for this wavelength.
19. Explain the technique of Doppler cooling? What is the temperature limit that can be attained through this method?
20. What is an optical molasses? What is its role in cooling atoms?
21. What is a magneto-optic trap? Why is it required in cooling of atoms?
22. Describe the technique of evaporative cooling.
23. Discuss the four-level (pumping) scheme for laser action.
24. Explain the principle of an atom laser and describe its important components.

25. Explain with neat diagram absorption, spontaneous emission and stimulated emission of radiation.
26. What is population inversion? Explain why laser action cannot occur without population inversion between atomic levels.
27. What do you understand by a negative temperature state? How can it be achieved?
28. Describe the working of solid state ruby laser.
29. What is meant by Bose-:Einstein condensation? What is its importance?
30. In helium -neon laser why is it necessary to use narrow tubes?
31. What is the reason for monochromaticity of laser beam?
32. Two if sources of light of intensities f and 6I are placed 200cm apart. At what place should a screen be placed on the line joining the sources, so that illumination due to the two sources may be equal on the screen.
33. For the-human eye, consider the pupil diameter ~3 mm, mean wavelength ~600A; and focal length ~20 mm. Compute the area of the image of an ideal point source at infinity as determined by diffraction. If the real density of the cones in the retina is designed by nature to use the optimum available resolution, estimate the area! density of the cones on the retina.
34. The angular diameter of the moon, as seen from f the earth, is 0.008 radian, and the moon's surface re-emits 20% of the light incident on it from the sun. Deduce the ratio of illumination on the earth in full moon light to that in full sunlight.